AGRO-INDUSTRIES FOR DEVELOPMENT

D0121017

AGRO-INDUSTRIES FOR DEVELOPMENT

Edited by

Carlos A. da Silva
Doyle Baker
Andrew W. Shepherd
Chakib Jenane

and

Sergio Miranda-da-Cruz

Published by

The Food and Agriculture Organization of
the United Nations

and

The United Nations Industrial Development Organization

by arrangement with

CAB International

www.cabi.org

CABI is a trading name of CAB International

CABI Head Office
Nosworthy Way
Wallingford
Oxfordshire OX10 8DE
UK

CABI North American Office
875 Massachusetts Avenue
7th Floor
Cambridge, MA 02139
USA

Tel: +44 (0)1491 832111
Fax: +44 (0)1491 833508
E-mail: cabi@cabi.org
Website: www.cabi.org

Tel: +1 617 395 4056
Fax: +1 617 354 6875
E-mail: cabi-nao@cabi.org

A catalogue record for this book is available from the British Library, London, UK.

Library of Congress Cataloging-in-Publication Data
Agro-industries for development / edited by Carlos A. da Silva ... [et al.].
    p. cm.
    Includes bibliographical references and index.
    ISBN 978-1-84593-576-4 (hardback : alk. paper) -- ISBN 978-1-84593-577-1 (pbk. : alk. paper)
1. Agricultural industries--Developing countries. 2. Agriculture--Economic aspects--Developing countries.
3. Economic development--Developing countries. 4. Competition--Developing countries.
I. Da Silva, Carlos A. II. Title.

HD9018.D44A39 2009
338.109172'4–dc22

2009006652

Published jointly by CAB International and FAO. Food and Agriculture Organization of the United Nations (FAO) Viale delle Terme di Caracalla, 00153 Rome, Italy. Website: www.fao.org

    ISBN: 978 1 84593 576 4 (CABI hardback edition)
    ISBN: 978 1 84593 577 1 (CABI paperback edition)
    ISBN: 978 92 5 106244 9 (FAO hardback edition)
    ISBN: 978 92 5 106019 9 (FAO paperback edition)

Typeset by SPi, Pondicherry, India.
Printed and bound in the UK by the MPG Books Group, Bodmin.

Contents

Contributors

José Miguel Aguilera, *Professor, Department of Chemical Engineering, Universidad Católica de Chile, Santiago, Chile*

Doyle Baker, *Chief, Rural Infrastructure and Agro-Industries Division, Food and Agriculture Organization of the United Nations, Rome, Italy*

Ralph Christy, *Professor, Department of Applied Economics and Management, Cornell University, Ithaca, New York, USA*

John Cranfield, *Associate Professor, Department of Food, Agricultural and Resource Economics, University of Guelph, Ontario, Canada*

Carlos A. da Silva, *Agribusiness Economist, Rural Infrastructure and Agro-Industries Division, Food and Agriculture Organization of the United Nations, Rome, Italy*

Colin Dennis, *Director General, Campden BRI, Chipping Campden, Gloucestershire, UK*

Claudia Genier, *Senior Consultant, FSG Social Impact Advisors, Geneva, Switzerland*

Spencer Henson, *Professor, Department of Food, Agricultural and Resource Economics, University of Guelph, Ontario, Canada*

Alain de Janvry, *Professor, Agriculture and Resource Economics, University of California at Berkeley, USA*

Mark Lundy, *Agroenterprise Specialist, International Center for Tropical Agriculture (CIAT), Cali, Colombia*

Edward Mabaya, *Researcher Associate, Department of Applied Economics and Management, Cornell University, Ithaca, New York, USA*

James MacGregor, *Researcher, International Institute for Environment and Development (IIED), London, UK*

Nomathemba Mhlanga, *PhD Candidate, Department of Applied Economics and Management, Cornell University, Ithaca, New York, USA*

Emelly Mutambatsere, *Evaluation Analyst, African Development Bank, Tunis, Tunisia*

Marc Pfitzer, *Managing Director, FSG Social Impact Advisors, Geneva, Switzerland*

Rudi Rocha, *PhD Candidate, Department of Economics, Pontifícia Universidade Católica do Rio de Janeiro, Brazil*

Morton Satin, *Director of Technical and Regulatory Affairs, Salt Institute, Alexandria, Virginia, USA*

Mike Stamp, *Consultant, FSG Social Impact Advisors, Geneva, Switzerland*

Bill Vorley, *Head, Sustainable Markets Group, International Institute for Environment and Development (IIED), London, UK*

John Wilkinson, *Professor and Researcher, CPDA, Universidade Federal Rural do Rio de Janeiro, Brazil*

Norbert Wilson, *Associate Professor, Department of Agricultural Economics, Auburn University, Auburn, Alabama, USA*

Foreword

Developing competitive agro-industries is crucial for generating employment and income opportunities. It also contributes to enhancing the quality of, and the demand for, farm products. Agro-industries have the potential to provide employment for the rural population not only in farming, but also in off-farm activities such as handling, packaging, processing, transporting and marketing of food and agricultural products. There are clear indications that agro-industries are having a significant global impact on economic development and poverty reduction, in both urban and rural communities. However, the full potential of agro-industries as an engine for economic development has not yet been realized in many developing countries, especially in Africa.

To address these issues, the Food and Agriculture Organization of the United Nations (FAO), the United Nations Industrial Development Organization (UNIDO) and the International Fund for Agricultural Development (IFAD) organized the first Global Agro-Industries Forum (GAIF) in New Delhi, India, from 8 to 11 April 2008. The Forum developed a shared vision on the factors critical to the future development of agro-industries, the key factors affecting their competitiveness, and potential priority action areas. The objectives of the Forum were threefold: to learn lessons from previous efforts and successes to develop competitive agro-industries in the developing world; to ensure stronger collaboration and joint activities among multi-lateral organizations working on agro-industrialization; and to clarify the distinctive roles of the public sector, multi-lateral organizations and the private sector in agro-industrial development. A related objective was to engage international organizations and financial institutions into launching initiatives at national and regional levels to foster agro-industrial development.

FAO, UNIDO and IFAD are committed partners for the development of a shared vision to maximize the impact of the agro-industrial sector on the livelihoods of those in the developing world. Our agencies are working together to assist their Member States in creating enabling environments for the development of agribusiness, agro-industries and agro-based value chains. We are doing this through the formulation and implementation of strategies for improving policies, regulatory frameworks, institutions and services. We are also promoting the incorporation of agro-industrial development strategies into country level programme frameworks and strategic action plans to assist the poor and small farmers.

This publication is an outcome of the Global Agro-Industries Forum. It has evolved through contributions from scholars and development practitioners aimed at highlighting the current status and future course of agro-industries and bringing further attention to the valuable contribution that the agro-industrial sector can make to international development. FAO, UNIDO and IFAD expect that the materials presented here will help advance the knowledge and enrich the debate on the role of agro-industries in generating employment, creating income, and fighting poverty in the developing world.

Jacques Diouf   **Kandeh K. Yumkella**   **Kanayo Nwanze**
Director-General   *Director-General*   *President*
FAO   *UNIDO*   *IFAD*

Acknowledgements

This book has been prepared through a collaborative effort of FAO and UNIDO, under the technical leadership of Carlos A. da Silva (Agribusiness Economist, Rural Infrastructure and Agro-Industries Division, FAO), Doyle Baker (Chief, Rural Infrastructure and Agro-Industries Division, FAO), Andrew W. Shepherd (Marketing Economist, Rural Infrastructure and Agro-Industries Division, FAO), Chakib Jenane (Chief of the Agro-Industries Branch, UNIDO) and Sergio Miranda-da-Cruz (Director, UNIDO Agribusiness Development Branch).

The editors wish to acknowledge the contributions made by the technical officers of FAO and UNIDO who participated in the conceptualization of the book chapters and in their review. Thanks are due to Roberto Cuevas-Garcia, Stephanie Gallat, Eva Gálvez, David Kahan, Danilo Mejia, Divine Njie, Rosa Rolle, Maria Pagura, Alexandra Röttger and Gavin Wall, from FAO, as well as to Karl Schebesta and Sean Peterson, from UNIDO.

We are also grateful to Geoffrey Mrema, Director of the Rural Infrastructure and Agro-Industries Division, FAO, for his support to this initiative.

The guidance of Rachel Tucker, from FAO's Electronic Publishing Policy and Support Branch, during the several stages of the book's publication process is greatly appreciated.

Finally, a special word of thanks is due to the chapter authors, for their commitment to this project and for their readiness to promptly react to the editorial comments.

Rome and Vienna,
May 2009
The Editors

1 Introduction

CARLOS A. DA SILVA[1] AND DOYLE BAKER[2]

[1]*Agribusiness Economist, Rural Infrastructure and Agro-Industries Division, FAO, Rome, Italy;* [2]*Chief, Rural Infrastructure and Agro-Industries Division, FAO, Rome, Italy*

The demand for food and agricultural products is changing in unprecedented ways. Increases in per capita incomes, higher urbanization and the growing numbers of women in the workforce engender greater demand for high-value commodities, processed products and ready-prepared foods. A clear trend exists towards diets that include more animal products such as fish, meat and dairy products, which in turn increases the demand for feed grains (FAO, 2007). There is also a growing use of agricultural products, particularly grains and oil crops, as bioenergy production feedstock. International trade and communications are accelerating changes in demand, leading to convergence of dietary patterns as well as growing interest in ethnic foods from specific geographical locations.

The nature and extent of the changing structure of agrifood demand offer unprecedented opportunities for diversification and value addition in agriculture, particularly in developing countries. As a reflection of changing consumer demand, the 1990s witnessed a diversification of production in developing countries into non-traditional fruits and vegetables. The share of developing countries in world trade of non-traditional fruits and vegetables has increased rapidly in the recent past (FAO, 2007). According to Rabobank, global processed foods sales per year are estimated at well over US$3 trillion, or approximately three-quarters of the total food sales internationally (Rabobank, 2008). While most of these sales are in high-income countries, the percentages of global manufacturing value addition for the main agro-industry manufacturing product categories generated by developing countries have nearly doubled in the last 25 years (FAO, 2007).

The prospects for continued growth in demand for value-added food and agricultural products constitute an incentive for increased attention to agro-industries development within the context of economic growth, food security and poverty-fighting strategies. Agro-industries, here understood as a component of the manufacturing sector where value is added to agricultural raw materials through processing and handling operations, are known to be efficient engines of growth and development. With their forward and backward linkages, agro-industries have high multiplier effects in terms of job creation and value addition. A new dairy processing plant, for instance, creates jobs not only at its own transformation facilities, but also at dairy farms and in milk collection, farm input supply and product distribution. The demand pull created by an agro-industrial enterprise stimulates businesses well beyond the closest links with its direct input suppliers and product buyers; a whole range of ancillary services and supporting activities in the secondary and tertiary sectors of the economy are also positively impacted. Because of the generally perishable and bulky characteristics of agricultural products, many agro-industrial plants and smaller-scale agro-processing enterprises tend to be located close to their major sources of raw materials. Consequently, their immediate socio-economic impacts tend to be exerted in rural areas.

The World Development Report 2008 (World Bank, 2007) called attention to the fact that some 800 million people are considered poor, subsisting with incomes of less than US$1 per day. Among the world's poor, 75% live in rural areas, having agriculture as a major source of livelihood. Fighting poverty will require that economic growth and development are brought to rural areas. Agro-industries, as will be argued in the ensuing chapters of this book, are part of the answer to this challenge.

The accelerated growth of agro-industries in developing countries also poses risks in terms of equity, sustainability and inclusiveness. Where there is unbalanced market power in agrifood chains, value addition and capture can be concentrated among one or a few chain participants, to the detriment of the others. Agro-industries will be sustainable only if they are competitive in terms of costs, prices, operational efficiencies, product offers and other associated parameters and only if the prices they are able to pay farmers are remunerative for those farmers. Establishing and maintaining competitiveness constitute a particular challenge for small- and medium-scale agro-industrial enterprises and smaller-scale farmers. Although agro-industries have the potential to provide a reliable and stable outlet for farm products, the need to ensure competitiveness favours farmers who are better able to deliver larger quantities and better quality of products. To the extent that smaller, resource-poor farmers are left out of supply chains, the socio-economic benefits of agro-industries are potentially reduced. A need thus exists for policies and strategies that, while promoting agro-industries, take into account issues of competitiveness, equity and inclusiveness.

The rapid rise in food prices witnessed in 2007 and 2008 was a stark reminder that the changing nature of agrifood systems, and how policy makers respond to the changes, can have immediate humanitarian and political consequences. Agricultural sector and agro-industry adjustments in the 1990s and the early 2000s contributed to reductions in the supply and international reserves of staple foods. The global food system did not have the capacity to respond to a 'perfect storm' of events that impacted on both short-term supply and demand. As political consequences of food price spikes and shortages mounted, policy responses that included export bans further worsened an already unbalanced market situation. The food prices crisis, though it has already subsided, points to the importance of the recent trends in agrifood systems, as well as the need for sound policies and strategies that enhance the competitiveness and developmental impact of agro-industries.

This book consists of a collection of readings that explore different elements of the broad issues associated with the development of agro-industries that are competitive, equitable and inclusive, with a focus on developing countries. The chapters were commissioned from a number of renowned scholars and development practitioners by the Food and Agriculture Organization of the United Nations (FAO), the United Nations Industrial Development Organization (UNIDO) and the International Fund for Agricultural Development (IFAD) to form the core of the technical programme of the Global Agro-Industries Forum, organized by these three agencies in April 2008 in New Delhi, India. The Global Agro-Industries Forum (GAIF) aimed to develop a shared vision of the factors critical to future developments of agro-industries, to learn from success stories in promoting competitive agro-industries in the developing world, to ensure stronger collaboration and joint activities among multilateral organizations working on agro-industrialization and to clarify the roles of the private sector, public sector and multilateral organizations in agro-industrial development. A further objective included the engagement of multilateral organizations and financial institutions in launching initiatives at national and regional levels to foster agro-industrial development.

The themes covered in this book were subjects for the Forum's plenary addresses. The plenary addresses were instrumental in calling attention to the status of agro-industries in the world, providing analytical insights on key trends and issues, considering future developments and assessing agro-industry policy issues and priorities. Following the Global Forum, the plenary addresses were further developed and are presented here as a sequence of six chapters. In addition, the keynote address of Professor Alain de Janvry, from the University of California at Berkeley, is presented as a special annex highlighting aspects of agribusiness and agro-industry development that were considered in the recent World Development Report 2008 (World Bank, 2007).

The three chapters following this introduction together provide an overview of the main trends, characteristics and impacts of agro-industries in developing countries. A cross-cutting theme in these chapters is the importance of viewing agro-industries within the context of the wider restructuring of agrifood systems. Agro-industrialization is not so much promoted as being seen as a consequence of external drivers. While there are marked differences among countries and regions with respect to the degree of structural and organizational transformation, the processes of agro-industrialization have widespread and profound impacts. The potential impacts are so significant that the processes must be understood and sound policy responses put in place to optimize potential benefits while mitigating risks. All three chapters provide insights into the challenges that policy responses need to address.

In Chapter 2, Spencer Henson and John Cranfield, from the University of Guelph in Canada, characterize the processes of agro-industrialization in developing countries and build a political case for agro-industries as a driver of growth and development. Henson and Cranfield develop their case around two main arguments. One cornerstone of their argument is that rapid changes in agrifood systems are shifting the basis for competitiveness. Increasingly, competitiveness is being determined by factors such as economies of scale, efficiencies in logistics, compliance with stringent grades and standards, and capacity to reach global markets with differentiated products. Henson and Cranfield observe that countries that have achieved higher integration with global markets with high-value products, or countries with large high-value domestic markets, seem to have advanced the most in terms of the contribution of agriculture to economic development. Their second argument relates to the pervasiveness of the impacts of agro-industries. They point to the key distributional consequences and discuss potential environmental consequences.

The main message of Henson and Cranfield is that countries must think and act strategically in order to cope with the challenges, starting with the important strategic choice on how countries and firms position themselves with respect to market competition. They stress that policy makers need to define their roles vis-à-vis the private sector and need to establish effective public–private working relations. Chapter 2 makes it clear that Henson and Cranfield believe that a key role for the public sector is to create conditions that allow the development of cost-competitive agro-industries. Some of the key challenges identified by the authors include improved infrastructure and access to finance, as well as macro-economic and trade policies that are conducive to investment and innovation.

The case for agro-industries development is reinforced in Chapter 3, written by John Wilkinson and Rudi Rocha, researchers from, respectively, the Federal Rural University of Rio de Janeiro and the Pontificate Catholic University of Rio de Janeiro, Brazil. Drawing from an extensive range of statistical data and empirical research sources, the chapter characterizes the contributions of agro-industries to economic development worldwide. Wilkinson and Rocha particularly emphasize contributions to manufacturing value addition and employment generation. They point out, however, that it is not possible to fully appreciate the importance and impacts of agro-industries because much of the value addition and employment is in the informal sector.

Chapter 3 identifies several structural factors in domestic and global markets that reinforce the importance of promoting agro-industries in developing countries. Wilkinson and Rocha present data showing the increasing importance of processed agricultural products in agricultural trade, including in South–South trade and as a percentage of the food imports by developing countries. The authors also discuss the recent expansion in markets for differentiated food products, including fair trade, organic and origin-based products. They acknowledge that focus on these and other non-traditional exports as a strategy for driving agro-industrial development seems appealing, but is likely to be hampered by market access restrictions, tariff escalation and compliance costs for meeting increasingly stringent standards established by private organizations and large-scale buyers.

Wilkinson and Rocha conclude that policies for agro-industry development should occupy a central position in government strategies. They caution though that government strategies must be oriented to market sustainability and be a component of broader social policies that also aim at food and nutritional security.

In Chapter 4, the last of the chapters that focus on trends, characteristics and impacts of agro-industries, Colin Dennis (Campden BRI, UK), José Aguilera (Catholic University of Chile) and Morton Satin (Salt Institute, USA) discuss technology developments and their implications for agro-industries. Dennis and colleagues recall several of the trends identified in the preceding chapters and explain how these are driving technological development. The main premise of their chapter is that organized food industries and chains are needed to meet changing consumer requirements and feed expanding urban populations, and that performance of agricultural and food industries in turn will be highly dependent on the increased and cost-effective application of existing technologies, as well as exploitation of new and innovative technologies.

The analysis by Dennis, Aguilera and Satin highlights two overarching challenges in technology development. First, driven by the changes in consumer demand and market requirements, technologies are needed that can ensure specific food traits (safety, quality, nutritional value, etc.) at all stages through the life cycle of the end product. Second, because food is moving over long distances, including internationally, there is a need for technologies and practices to ensure the safety and quality of products for long periods. The authors argue that there will be an increasing need to meet sanitary and phytosanitary standards, and to complement technology development with development of effective food safety management systems.

The last three chapters turn to the critically important issue of the roles, responsibilities and actions of public and private sector actors in agro-industries development. A consistent theme in these chapters, indeed throughout the book, is that governments have an essential and legitimate role to play. At the same time, the message of these chapters is that agro-industries development is essentially a private sector activity. In the light of the developmental trends, challenges, benefits and risks highlighted in the first set of chapters, governments cannot be passive observers but they also should not attempt to control all aspects of agro-industries. Clarification of the roles and responsibilities of the public and private sectors in agro-industries development is one of the keys to achieving improved competitiveness and developmental impacts.

Enabling environments for competitive agro-industries are discussed in Chapter 5, written by Professor Ralph Christy of Cornell University and a team of collaborators. They argue that fostering competitive agro-industries requires that conducive business climates, or enabling environments, are in place. They recall recent efforts to promote reform processes through business climate assessments, but conclude that these approaches were not designed for the evaluation of business climates for agro-industrial enterprises.

To provide guidance on agro-industry-focused analyses of business climates, Christy and co-authors propose a hierarchy of state actions for characterizing and assessing enabling environments for agro-industrial enterprises, classifying actions as essential enablers, important enablers and useful enablers.

The hierarchy, however, is just a starting point. Christy and colleagues argue that, for effective reform to emerge, a nuanced appreciation of the roles that public policy makers can play in sustaining competitiveness is needed. The authors introduce and illustrate an analytical framework for reform processes framed by two key dimensions, namely the level of risk and uncertainty agro-industries face when conducting business and the capacity of the state in shaping the environment for business. The authors propose that the framework can be used to identify suitable policy options for different enabling environment reform contexts.

The issue of inclusiveness in agro-industries is the focus of Chapter 6, 'Business Models That Are Inclusive of Small Farmers', prepared by Bill Vorley, from the International Institute for the Environment and Development (IIED, UK), Mark Lundy, from the International Center for Tropical Agriculture (CIAT, Colombia) and James MacGregor, also from IIED. The authors define 'business model' as the way by which a business creates and captures value within a market network of producers, suppliers and consumers. The chapter describes a range of business models that improve the inclusiveness, fairness, durability and financial sustainability of trading relationships between small farmers and downstream agribusiness (processors, exporters and retailers).

Vorley and co-authors argue that the chief challenge for modern agrifood businesses in working with small-scale farmers is the difficulty of organizing supply chains so as to ensure that the benefits of logistics, economies of scale, traceability and compliance with private sector standards are achieved. They also contend that despite the difficulties faced, there are sound business reasons for agro-processors, retailers, exporters and other buyers to include small farmers in the farm-to-consumer value chain. The authors then introduce and illustrate a typology of organizational models, covering models organized by the producers themselves, by the end-customer companies or by an intermediary such as a trader, wholesaler or exporter. They argue that evidence on benefits and impacts of the different models is still weak, and that no single modality is inherently superior for smallholders.

Vorley and co-authors point out that despite the recent trends towards increased inclusiveness, the participation of smallholders and SMEs in modern markets is still more of an exception than the rule. They identify three priorities for enhancing competitiveness and inclusiveness of smaller-scale suppliers. The first is skills development to prepare farmers to be reliable partners and suppliers. The second, returning to the theme addressed by Christy and co-authors, is business-enabling environments. Vorley and colleagues emphasize the provision of key infrastructure services, public investments in services such as agricultural research, education and extension, and policies to maintain competitive markets. The third is for private sector actors to ensure that their procurement practices work to the benefit, rather than the detriment, of small-scale producers and suppliers. Vorley and co-authors give several examples of responsible business practices that can work to the benefit of small-scale suppliers.

The issue of responsible business practices is the subject of Chapter 7, prepared by Claudia Genier, Mike Stamp and Marc Pfitzer, from FSG Social

Impact Advisors, Switzerland. The chapter focuses on the concept of corporate social responsibility (CSR): what CSR means and how it has evolved over the past decade. The main theme of the chapter is that CSR has become for many a core business strategy oriented towards competitive advantage, partnerships along the supply chain, institution building and long-term sustainability. Under this new perspective, CSR strategies have the potential to increase the inclusiveness and competitiveness of agro-industries, creating a more equitable distribution of benefits along the value chain.

The authors characterize and assess various CSR codes and standards operating in agricultural value chains. They argue that standards and codes have helped to improve the quality, safety and traceability of food, but there is insufficient information to conclude that codes have improved environmental, social and economic conditions for producers. To the contrary, Genier and co-authors express concern that the proliferation of standards and codes, as well as high implementation costs, can lead to the marginalization of small producers.

One of the main contributions of Genier and co-authors is to expand the scope of what is generally considered to be corporate responsibility. They point out that more visionary agrifood companies – recognizing the drawbacks of reliance on standards and codes – have adopted value chain innovations that seek to expand economic opportunities along entire chains. The authors review several cases of value chain innovations and conclude that evidence of impact can be found in the various initiatives in terms of quality, health and safety improvements, better environmental indicators, higher productivity and development impacts. However, they caution that the cases they appraised remain a minor exception in comparison with the core business practices of many agrifood industries, and argue that governments and civil society have an important role to play in scaling up and replicating value chain innovations.

All six chapters address in one way or another the fundamental policy dilemma of agro-industries development: the need to establish and maintain competitiveness while also addressing the risks to smaller-scale economic actors. The authors of these chapters do not view policy support to agro-industries as a choice between competitiveness and developmental impacts, but rather see it as essential for enhancing both. One of the important contributions of the following chapters is to clarify the challenges being faced and identify strategies and practical actions to address them.

There are several messages about agro-industries development that cut across the chapters of this book. One is that governments clearly do have an important role to play. To enhance competitiveness, enabling policies and institutions must be put in place and infrastructure must be improved, particularly rural infrastructure. Recommendations on other specific priorities for establishing enabling environments are made by most of the authors.

Another theme found in all the chapters is that agro-industry firms and value chain stakeholders must be ready to meet the challenges of changing consumer requirements and market competition. Priority attention should be given to consumer concerns and interests regarding quality, safety, health benefits, product origin and other attributes. To access higher-valued markets, capacity is needed to develop, distinguish and certify specific product traits. There is also a need to improve productivity and efficiency. Systematic attention is required to build capacity for acquiring and utilizing productivity-enhancing technologies. The capacity to introduce and apply advanced techniques for supply chain management and logistics will increasingly become a requirement for competitiveness of agro-industries targeting global and regional markets.

An important theme, particularly stressed by Vorley and co-authors and Genier and co-authors, is that value chains that include smaller-scale producers and processors can make good business sense. There nevertheless are many reasons why firms choose not to work with smaller-scale suppliers. To achieve objectives relating to economic growth and rural development, public and private sector initiatives are needed to strengthen business linkages and support the development of business models that include smaller-scale producers and processors. The development of inclusive business models requires, in turn, concerted efforts to organize smallholders and build the capacities of farmers to be reliable suppliers. Financial services and products that fit the specific conditions of producers, processors and others in the supply chain are also critical for achieving widespread developmental impacts.

While the authors present a consistent and coherent overview of agro-industry drivers, trends, challenges and responses, all are careful to point out that there is great diversity in circumstances. There is a corresponding need to ensure that policies and strategies to improve competitiveness and developmental impacts are based on a solid understanding of broader market, consumer and technological trends, as well as the specific conditions of each country, agro-industry and agricultural value chain.

Finally, it is worth recalling that agro-industries development is such an all-inclusive and complex process that not all issues could be comprehensively addressed adequately in a single book or during the GAIF. Several such issues are briefly touched on in the following chapters, even if not central to the message of any single chapter. One issue is the growing urgency to consider whether and how to work towards the harmonization of national and international regulatory frameworks. Another issue is the importance of the informal sector in agro-processing. Environmental consequences of agro-industries are discussed by Henson and Cranfield but, overall, this volume does not emphasize this important topic. The chapters do present a convincing case on the importance of agro-industries in developing countries, and point to some of the policy priorities for enhancing competitiveness and developmental impacts. However, this is just a starting point. Issues such as those just identified make it clear that there remains a great need for additional analysis of trends and policy responses. FAO, UNIDO and IFAD, the United Nations agencies that organized the GAIF and the contributions to the present book, are working to fill these gaps, and are committed to promoting international agro-industrial development that is sustainable, inclusive and equitable.

References

FAO (2007). Challenges of Agribusiness and Agro-industries Development, Committee of Agriculture, Twentieth Session, COAG/2007/5, Rome, Italy.

Rabobank (2008). The Boom Beyond Commodities: A New Era Shaping Global Food and Agribusiness, Hong Kong.

World Bank (2007). World Development Report 2008 – Agriculture for Development, Washington, DC.

2

Building the Political Case for Agro-industries and Agribusiness in Developing Countries

SPENCER HENSON[1] AND JOHN CRANFIELD[2]

[1]*Professor, Department of Food, Agricultural and Resource Economics, University of Guelph, Ontario, Canada;* [2]*Associate Professor, Department of Food, Agricultural and Resource Economics, University of Guelph, Ontario, Canada*

Introduction

One of the most profound changes taking place in the agro-food economy of developing countries is the emergence of agro-industrial enterprises as part of broader processes of agribusiness development. In turn, the transformation of agro-processing from the informal to the formal sector has critical implications for participants along the entire length of the supply chain, from those engaged in agriculture, fisheries and forestry through food retailers and traders to the final consumer. Potentially, agro-industrialization presents valuable opportunities and benefits for developing countries, in terms of overall processes of industrialization and economic development, export performance, food safety and quality. At the same time, however, there are potentially adverse effects on those engaged in informal sector agro-processing enterprises, such that processes of agro-industrialization must be attuned with overall processes of economic restructuring. Further, agro-industries are changing on a global scale, presenting not only new opportunities but also challenges for developing countries, and suggesting that the future trajectory of agro-industrialization will be somewhat different than in the past.

The aim of this chapter is to explore the political case for agro-industrialization in developing countries, highlighting both the likely benefits and the areas where caution is needed, and where critical actions can steer this process along the most beneficial path. In so doing, the chapter addresses four key questions:

- What are the characteristics of the agro-industrial sector?
- How are the processes of agro-industrialization proceeding and what are driving these?

- What impact is agro-industrialization having on developing countries?
- What are the challenges for developing countries in promoting agro-industrialization in a manner that is of maximum benefit?

Examples and data are provided to illustrate key points. The chapter concludes by suggesting notable areas where action is required to ensure that processes of agro-industrialization proceed unimpeded in developing countries and in a manner that makes the maximum contribution to overall processes of industrialization and economic development.

Nature of the Agro-industrial Sector

The agro-industrial sector is here defined as the subset of the manufacturing sector that processes raw materials and intermediate products derived from agriculture, fisheries and forestry. Thus, the agro-industrial sector is taken to include manufacturers of food, beverages and tobacco, textiles and clothing, wood products and furniture, paper, paper products and printing, and rubber and rubber products, as in FAO (1997). In turn, agro-industry forms part of the broader concept of agribusiness that includes suppliers of inputs to the agricultural, fisheries and forestry sectors and distributors of food and non-food outputs from agro-industry.

Most agricultural, fisheries and forestry production is subject to some form of transformation beyond the farm gate and prior to eventual end use. From the outset this transformation highlights the critical role that agro-industry plays in supply chains. At the same time, the roles that agro-industry is playing are changing over time, and the distinction from other sectors is becoming less clear as technologies cut across industries (e.g. biotechnology). Moreover, agro-industries are increasingly using inputs that they have not traditionally employed, while non-agro-industries are beginning to use raw materials from agriculture, fisheries and forestry.

The key defining characteristic of the agro-industrial sector is the perishable nature of the raw materials it employs, the supply and/or quality of which can vary significantly over time. Under conditions of uncertain raw material supply, planning of production and transformation processes and achieving economies of scale can be fraught with difficulty, especially where there are very specific quality parameters (e.g. canning of fruits and vegetables). Thus, there is an incentive for agro-industries to engage in primary production themselves (as with plantation systems) or to develop longer-term supply relations with producers, aimed at improving efficiency in production, securing a reliable supply, promoting the adoption of varieties that are best suited to processing operations, and so on.

The manufacture of food products, especially in the context of a developing country, typically involves a relatively limited range of technologies that do not differ widely across product categories. In most cases the level of value added is relatively limited, such that raw materials account for a significant proportion of end-product prices. In contrast, a great variety of raw materials are used in the manufacture of non-food agro-industrial products, while there are diverse product end uses. The level of transformation undertaken in the non-food agro-industrial sector is usually considerable, such that the level of value added is high and raw materials account for a smaller proportion of the end-product price. Furthermore, a wide variety of technologies are typically employed both within and across non-food agro-industrial product categories. Across both the food and non-food agro-industrial sub-sectors, however, there is a trend towards greater levels of transformation and value addition, and the employment of more advanced technologies.

While recognizing the broad characteristics of food and non-food agro-industries in developing countries noted above, the processes involved can vary from artisan or craft skills to industrial processes, across the informal and formal sectors. Indeed, within any one product sub-sector (e.g. grain milling or the manufacture of paper) it is possible to observe diverse technologies operating side by side. Further, there may be significant interlinkages between enterprises that employ low levels of technology, predominantly in the informal sector, and those employing more advanced technologies, predominantly in the formal sector. Examples include the subcontracting of particular functions and/or the handling of by-products and waste from manufacturing processes. This suggests potentially significant and complex relations between different forms of business and across the informal and formal sectors, which may also link agro-industries to other sectors.

The coexistence of the informal and formal sectors is perhaps one of the key distinguishing features of the agro-industrial sector in developing countries. While national accounts of most countries largely ignore the economic activities of the informal sector, in most low-income countries informal or local agro-processing remains strong. Indeed, 'informality' can be considered the norm in the agro-industrial sector, with formal sector enterprises comprising a relatively small fraction of the use of agricultural, fisheries and forestry raw materials, value addition and employment (Sautier et al., 2006). The informal sector itself, however, represents a highly transitory collection of enterprises, with rates of firm closure averaging between 9% and 10% annually (Mead, 1994; Mead and Liedholm, 1998). Indeed, in many cases it is difficult to even conceptualize informal agro-processing activities as 'enterprises'; often individuals may be involved in multiple business activities that can change from season to season, and even from hour to hour during the day. As we will see, rather than representing the establishment of new enterprises and industries, the development of the formal agro-industrial sector represents a transition from 'informality' to 'formality' as the predominant business form and mode of industry organization. The economic consequences of the evolution of agro-industries in developing countries need to be viewed in this context.

Evolution of the Agro-industrial Sector

Since the early 1990s, there has been a rapid process of agro-industrialization in many developing countries, characterized by the establishment of private and formal sector firms across an increasing array of food and non-food sectors. In order to understand the nature and consequences of this evolution, however, it needs to be viewed in the context of the wider restructuring of the entire agribusiness complex. In this regard we can posit three broad sets of changes (Reardon, 2007). First, the growth of agro-processing, distribution and agricultural input provision activities off-farm by agro-industrial firms. Second, institutional and/or organizational changes to the relations between agro-industrial firms and primary producers, for example, increasing levels of vertical integration. Third, changes in the primary production sector in terms of product composition, technology, sectoral and market structures, etc. (Reardon and Barrett, 2000). Thus, we can see the growth of the agro-industrial sector as being integral to profound changes in the entire way in which the agro-food complex is structured and organized. In turn, this suggests impacts on actors at all levels of the supply chain, from primary production to consumption. The framework developed by Reardon and Barrett (2000) provides a useful lens through which one can understand these processes of agro-industrialization in developing countries, the factors driving these processes and their consequences (Figure 1).

Underlying meta-trends

Underlying the evolution of the agro-industrial sector is a broad set of meta-trends, at both the national and international levels, that condition the way in which the sector is structured and operates over time. With respect to domestic markets for the products of agro-industries, population and income growth are driving changes in food consumption patterns at the broad commodity level, away from starchy staples and towards meat, dairy products, fruits and vegetables, oils and processed grains (see e.g. Cranfield *et al.*, 1998; Pingali and Khwaja, 2004), reflecting the patterns predicted by Bennett's law[1] (Table 1). With increasing urbanization (Figure 2), greater participation of women in the paid labour force and greater ownership of household appliances (e.g. refrigerators and microwave ovens), demand for more highly processed and higher-value food products with high income elasticities is growing (Figures 3 and 4). This trend is driving the evolution of the food-processing sector and providing a mechanism through which enterprises can counteract the downwards pull on relative food expenditure exerted by Engel's law.[2] In turn, this leads to an increased demand for raw materials from primary production, accompanied by shifts in the types and qualities of raw materials being demanded, which can generate economic benefits for the agriculture, fisheries and forestry sectors (Reardon and Barrett, 2000).

Through the 1980s and 1990s, the political economy in which agro-industries operated changed radically, both nationally in developing countries and internationally. Agro-industries shifted operations from a predominantly

---

[1] Editors' notes: Bennett's law posits that, as income rises, per capita consumption of starchy food staples falls.
[2] Engel's law states that, as incomes increase, the proportion of income spent on food falls.

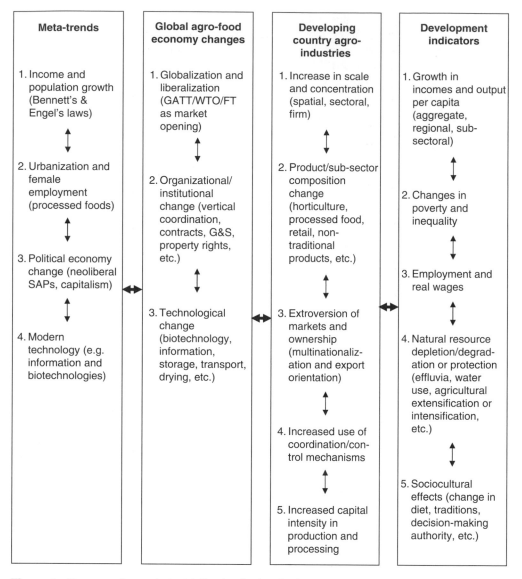

Meta-trends	Global agro-food economy changes	Developing country agro-industries	Development indicators
1. Income and population growth (Bennett's & Engel's laws)	1. Globalization and liberalization (GATT/WTO/FT as market opening)	1. Increase in scale and concentration (spatial, sectoral, firm)	1. Growth in incomes and output per capita (aggregate, regional, sub-sectoral)
2. Urbanization and female employment (processed foods)	2. Organizational/institutional change (vertical coordination, contracts, G&S, property rights, etc.)	2. Product/sub-sector composition change (horticulture, processed food, retail, non-traditional products, etc.)	2. Changes in poverty and inequality
3. Political economy change (neoliberal SAPs, capitalism)	3. Technological change (biotechnology, information, storage, transport, drying, etc.)	3. Extroversion of markets and ownership (multinationalization and export orientation)	3. Employment and real wages
4. Modern technology (e.g. information and biotechnologies)		4. Increased use of coordination/control mechanisms	4. Natural resource depletion/degradation or protection (effluvia, water use, agricultural extensification or intensification, etc.)
		5. Increased capital intensity in production and processing	5. Sociocultural effects (change in diet, traditions, decision-making authority, etc.)

Figure 1. Process of agro-industrialization in developing countries. (From Reardon and Barrett, 2000.)

statist model, through structural adjustment and market liberalization, to a focus on the private sector and establishing conditions that encourage private entrepreneurial behaviour. This shift has arguably enhanced the opportunities for private investment in the agro-industrial sector and reduced the costs of cross-border flows of both goods and capital (Reardon and Barrett, 2000).

We are also observing technological advances both generally (most notably information and communication technologies) and, specific to the agro-industrial sector, in primary production (e.g. the application of biotechnology)

Table 1. Per capita consumption of food commodities in developing countries by region, 1970 and 2003. (From FAOSTAT; http://faostat.fao.org)

Food	Sub-Saharan Africa			Developing Asia			Latin America & Caribbean		
	1970	2003	% Change	1970	2003	% Change	1970	2003	% Change
Cereals (excluding beer)	110.9	123.3	11.1	155.2	163.3	5.2	114.7	129.8	13.1
Starchy roots	164.9	159.7	-3.2	64.3	45.7	-28.8	80.5	52.0	-35.5
Sugar and sweeteners	7.8	11.0	40.5	11.3	17.2	51.9	39.0	47.4	21.7
Pulses	11.3	9.6	-15.2	8.7	5.7	-34.1	13.6	11.8	-13.1
Tree nuts	1.3	0.9	-30.5	0.43	1.1	165.1	0.5	0.7	36.0
Vegetable oils	5.9	7.6	29.0	3.0	10.1	237.0	6.1	11.4	85.5
Vegetables	31.9	32.4	1.6	45.4	143.2	215.5	35.4	51.0	44.2
Fruits (excluding wine)	58.3	48.3	-17.2	21.4	49.0	128.6	93.8	98.9	5.5
Alcoholic beverages	42.2	34.9	-17.2	2.3	13.0	471.4	30.7	42.2	37.5
Meat	12.8	11.4	-11.0	7.6	27.8	265.6	34.7	59.1	70.4
Milk (excluding butter)	29.9	29.5	-1.2	20.0	42.4	112.5	84.6	105.9	25.1
Eggs	1.1	1.3	18.7	1.6	8.7	448.4	4.4	8.0	80.9
Fish, seafood	7.9	6.9	-12.8	6.1	16.1	162.9	7.0	8.5	22.0

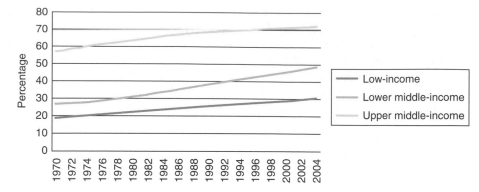

Figure 2. Proportion of population in urban areas by country income group, 1990–2004. (From World Bank World Development Indicators.)

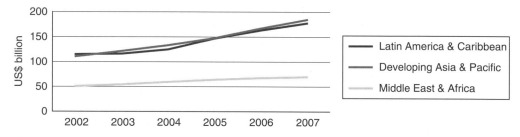

Figure 3. Sales of packaged foods in developing regions, 2002–2007 (US$ billion). (From Euromonitor data; www.euromonitor.com)

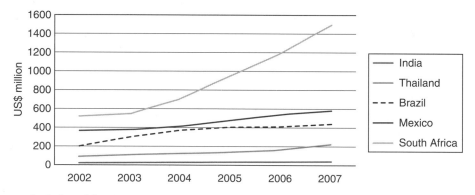

Figure 4. Sales of frozen processed foods in selected developing countries, 2002–2007 (US$ million). (From Euromonitor data; www.euromonitor.com)

and manufacturing sectors (e.g. new processing methods). These technological advances are serving to create new and unprecedented opportunities for agro-industrial enterprises in terms of product and process innovations, vertical and horizontal linkages within supply chains, the functioning of distributional systems, and so on. However, they also raise the spectre of agro-industrial enterprises being 'left behind' if they are unable to gain access to these technologies in a timely and cost-effective manner.

Alongside changes in domestic demand patterns in developing countries, shifts in consumption patterns in industrialized countries present potentially lucrative opportunities for developing country agro-industries, through higher-value exports. Examples include year-round demand for fresh and semi-processed fruits and vegetables for which developing countries have an agro-climatic advantage, and chilled and frozen fish and fishery products. At the same time, however, consumers in such markets are demanding greater assurances over food safety and quality, which require investments in more advanced systems of control through the supply chain.

Changes to the global agro-food economy

En masse, these meta-trends are fostering fundamental changes in agro-food systems, enhancing productivity and reducing transaction costs and fostering new modes of competitiveness within and across sectors. For example, the liberalization of global trade through the World Trade Organization (WTO), bilateral trade agreements and the preferential market access arrangements offered to low-income countries (in particular) has opened up higher-value industrialized country markets to developing country agro-industrial enterprises. Further, growth in domestic demand for processed foods provides an alternative route towards value addition for agro-industrial enterprises. At the same time, however, more liberal trade in agro-food products is also exposing domestic agro-industrial enterprises in developing countries to enhanced competition, both domestically and internationally. In addition, the challenges associated with this 'new reality', including technological innovation, increased scale of operation, coordination of activities vertically and horizontally and new institutional forms of governance, such as food safety and quality standards, intellectual property rights and contracting (see e.g. Reardon and Barrett, 2000; Henson and Reardon, 2005), require fundamental changes in the organization and conduct of agro-industrial enterprises and their economic relations with other parts of the agro-food system. It is far from certain that the successful businesses of the future will be the 'winners' from these changes; rather we may see the emergence of new enterprises that have the competencies required to compete in this more dynamic and liberal world.

In the case of agro-processed products, we would argue that one of the most fundamental changes in the governance of supply chains is the increasing role of grades and standards. In particular, dominant firms are utilizing product quality attributes as a means of product differentiation and market positioning (Raikes *et al.*, 2000; Busch and Bain, 2004). Indeed, it is argued that the very ways in which agro-food markets are structured and operate are increasingly defined by quality-based competition, while the associated institutional arrangements, both within and outside the supply chain, are crucial to the legitimacy of the quality attributes embedded in agricultural and food products (Allaire and

Boyer, 1995; Busch and Bain, 2004; Ponte and Gibbon, 2005; Busch and Bingen, 2006; Henson, 2007a). The credence nature of many of these attributes, including the impact of production processes on the environment, worker welfare, etc., contrasts to the predominant focus on search characteristics in most traditional markets.

In turn, the increasing focus on product safety and quality attributes has served to enhance the role of formal product and process standards. Standards are ubiquitous in market economies and play a fundamental role in the organization of supply chains for most products and services (Busch, 2000; Henson and Reardon, 2005), including traditional marketing systems. In higher-value markets, however, formal standards are increasingly the predominant mechanism of market and supply chain governance, operating alongside a plethora of non-codified buyer requirements (e.g. standards related to logistics). Indeed, standards are often a key vehicle for product differentiation in such markets (Henson and Reardon, 2005; Henson, 2007b). They also function as instruments of risk management by standardizing product requirements over suppliers, acting to reduce the transaction costs and risks associated with procurement, in particular where high levels of oversight are required to ensure food safety and/or quality attributes are delivered. Thus, we see leading buyers in supply chains (e.g. supermarket chains in industrialized countries) establishing standards on both an individual (e.g. Tesco Nature's Choice) and collective basis (e.g. GLOBALGAP[3] and the BRC Global Standards), alongside public and quasi-public national and international standards (e.g. ISO 9000) (Figure 5). Increasingly, certification to such a standard, or set of standards, is the minimum entry requirement for higher-value markets for agro-food products, not only in industrialized countries but also in developing countries' higher-income markets.

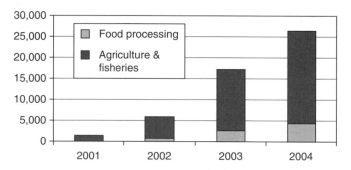

Figure 5. Global certifications to ISO 9000: 2000 in agriculture, fisheries and food processing, 2000–2004. (From ISO, 2005.)

---

[3] Editors' note: GLOBALGAP is a private sector body that sets voluntary standards for the certification of agricultural products internationally. The BRC Global Standards were established by the British Retail Consortium to set out quality and safety requirements for products supplied to their membership. ISO 9000 refers to a set of quality management standards specified by the International Organization for Standardization (ISO).

Restructuring of agro-industries in developing countries

The evolution of the agro-industrial sector is both a response to and an agent of the induced institutional and technological changes described above (Reardon and Barrett, 2000). This occurs through shifts in relative product and factor prices, enhanced flows of capital, transfers of technology and the evolution of organizational structures and institutions between enterprises and/or across sectors, etc. More broadly, while retaining distinct features related to the nature of agricultural products, for example, perishability and protracted production cycles, the agro-industrial sector in developing countries is evolving in a similar manner to commodity chains on a global basis (see e.g. Busch, 2000; Reardon *et al.*, 2003; Busch and Bain, 2004; Fold and Pritchard, 2005; World Bank, 2005). Thus, supply chains are increasingly extending beyond national and regional boundaries, facilitated in part by new food, communication and transportation technologies and a policy environment that encourages more liberal international trade (Nadvi and Waltring, 2003; OECD, 2004; Henson and Reardon, 2005). This is being accompanied by spatial agglomeration and firm concentration in agro-processing, such that a diminishing number of key economic players have power over global agricultural and food markets (Cook and Chaddad, 2000; Reardon and Barrett, 2000; Viciani *et al.*, 2001; Regmi and Gehlar, 2005), driving a shift towards buyer-driven supply chains for many products that are extending internationally with global sourcing and the emergence of multinational actors (Gereffi, 1999; Humphrey and Schmitz, 2001, 2003; Gereffi *et al.*, 2003).

Perhaps the most immediately evident trend in the agro-industrial sector in developing countries is a shift from the informal to formal sectors and, concurrently, increased concentration. Similar trends are observed in primary production, most notably in commodities for which there are potentially significant economies of scale (e.g. forestry and plantation crops). In many developing countries, agro-industry developed within the framework of existing institutional settings, plant cropping systems and marketing arrangements and norms that often pre-dated independence (Jaffee and Morton, 1995; Swinnen and Maertens, 2007). Large-scale and formal agro-industry was driven by the public sector in the form of parastatals, some of which were formed out of nationalized private enterprises, for example, in grain milling, vegetable canning and oil palm processing, which operated within a system of state-controlled commodity and input distribution. The underlying intention was to create economies of scale in production, shield farmers from market risks, overcome perceived gaps in entrepreneurship and force the transfer of resources from agriculture to nascent industrial sectors (Jaffee and Morton, 1995), often with donor support. This meant that the agro-industrial sector was frequently shielded from the competitive pressures that drove consolidation in other parts of the world.

Many state enterprises in the agro-industrial sector did not flourish for a whole host of reasons. These included political interference in the management of such enterprises in the pursuit of non-commercial objectives, bureaucratic burdens, development of structural deficits under conditions of managed input supply and output prices, and inadequate investment and access to new technologies. Through the 1980s, there was a fundamental shift in 'development thinking' regarding the principal roles of the public and private sectors that sparked off a radical process of structural reforms. Embedded in these reforms were efforts to restructure agricultural product and input marketing systems. In the 1990s, reforms extended to the privatization of state-owned agro-processing enterprises alongside the liberalization of markets for agricultural and other primary products. The efficacy of these reforms has, however, been mixed. In many cases state-owned enterprises were privatized under weak regulatory environments (Jaffee and Morton, 1995), while the slow evolution of financial markets and ancillary service providers acted to limit access to finance and inputs, and curtailed efforts to enhance efficiency through investments in new technologies. Further, processes of privatization were often protracted and often did not serve to transfer ownership on the basis of commercial criteria alone. In some cases, the privatization of parastatals has meant that they have been transferred as single entities into the private sector, while in others they have been disassembled into smaller operating units prior to sale or simply withered away, such that a less concentrated market structure has emerged. Nevertheless, enterprises must now survive in a more liberal environment where there is a definite thrust towards increased market concentration.

Alongside structural changes in the agro-industrial sector, trends in consumer demand and technological innovation are bringing about shifts in product composition (Reardon and Barrett, 2000) at both the primary commodity and processed product levels, towards sub-sectors where developing countries have a domestic and/or international competitive advantage. At the same time, the basis of competitiveness in markets for agro-industrial products is shifting, threatening the traditional areas of comparative advantage commanded by developing countries, but also providing new opportunities for enterprises that have access to the requisite capacities and resources. Broad thrusts in this shift are towards higher levels of value addition within the agro-industrial sector and away from traditional commodities and towards non-traditional crops and livestock products (e.g. fish, fresh fruits and vegetables, spices).

Broadly, agro-industries in developing countries have been traditionally based on the utilization of voluminous inputs that have relatively low unit values, but which are costly to transport. Thus, enterprises within agro-industries tended to be situated in close proximity to sources of raw materials. This contrasts with industries that are less tied to reliable supplies of unprocessed biological materials, including those that utilize primary processed agricultural, fisheries and forestry products as inputs into more highly processed products, which typically locate close to significant markets where the supply of inputs and capital tends to be more efficient, infrastructure is better, and so on. Indeed, most agro-industries in developing countries have evolved on the basis of strategic sources of raw materials; the palm oil sector in parts of sub-Saharan Africa and Asia is a good example. However, with the shift towards higher-value products for export and/or domestic markets, where the cost of raw materials represents a smaller proportion of the end-product price and/or where additional and a wider range of inputs are required (e.g. packaging and synthetic additives), some of which need to be imported, the advantages of being located on the basis of raw material supply is less apparent. Thus, we are seeing a new 'model' in terms of the location and competitive position of agro-industries in developing countries, which in most cases operates alongside traditional food and non-food agro-industrial sectors.

A further critical factor in the historic competitiveness of agro-industries in developing countries has been labour costs. Access to a plentiful supply of cheap labour explains, at least in part, why agro-processing enterprises in developing countries tend to be less capitalized than their industrialized country counterparts. While labour costs remain a key element of the competitiveness of some sub-sectors, for example, the production of semi-prepared fresh vegetables in Kenya for export to the European Union, in other sectors capacity utilization and the ability to meet consumer demands for product safety and quality are more critical, as in the dairy processing sector. Thus, also in Kenya, we have seen the emergence of new private sector firms that manufacture value-added dairy products aimed at particular market niches, including yogurts containing probiotics. The competitiveness of these enterprises bears little relation to the level of labour costs.

In many traditional agro-industries, especially those employing capital-intensive technologies, there are significant economies of scale that require operation at or near capacity on a continuous basis. This requires a reliable supply of raw materials, which is readily impeded by significant fluctuations in agricultural production and/or weak transport infrastructure, such that operation below capacity is often the norm rather than the exception. There is also a need for access to sizeable markets that may well exceed domestic demand, especially in a developing country context where per capita incomes are low and demand for more highly processed food products is only just emerging. Thus, agro-industries in developing countries are increasingly facing fierce competition from global enterprises rather than their regional counterparts. Countering this trend, however, the establishment of sizeable markets for higher-value agro-industrial products domestically and/or regionally, coupled with improvements in basic infrastructure, is providing opportunities for agro-industrial enterprises in developing countries to compete.

With improved infrastructure, the development of domestic markets for higher-value food and non-food products, access to improved technologies and improvements in labour productivity, more advanced agro-industrial enterprises are evolving in certain developing countries. Thus, we are observing increases in capital/labour ratios such that the competitive position of these enterprises is less reliant on proximity to supplies of raw materials and/or low labour costs than is the case with more traditional agro-industrial enterprises. Such enterprises may be directed at evolving domestic and/or regional markets, or at export markets in industrialized countries, spurred by access to trade preferences and broader processes of trade liberalization.

In many developing countries the transformation of the agro-industrial sector has involved, and in some cases been fundamentally driven by, foreign direct investment (FDI), predominantly by multinational corporations. Such investments have taken the form of acquisitions of (or mergers with) existing domestic enterprises, of joint ventures or of the establishment of new enterprises. While foreign investment is nothing new in many developing countries, having been a common characteristic of the plantation-processing sector (e.g. involving Unilever and Del Monte), multinationals are now investing in stand-alone processing operations, often directed at domestic and regional markets (e.g. Nestlé and Coca Cola) or the grocery retail sector (e.g. Wal-Mart, Carrefour and Tesco). In part, such investments reflect an overall trend towards enhanced flows of FDI (Figure 6) as corporations in industrialized countries look for investments that have the potential to yield greater returns. Thus, FDI flows to developing countries, and especially Asia, have grown rapidly over the last 10–15 years (Figure 7). At the same time, FDI flows to the food manufacturing sector have also increased dramatically (Figure 8), as multinationals are drawn to the opportunities presented by rapidly expanding markets for higher-value food

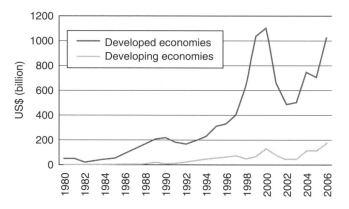

Figure 6. Value of foreign direct investment (FDI) from developed and developing economies, 1980–2005. (From UNCTAD Foreign Direct Investment Database.)

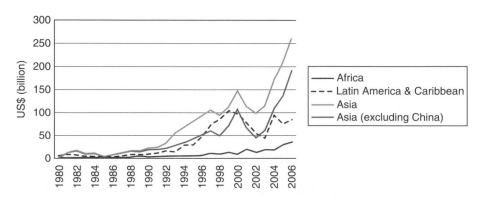

Figure 7. Value of foreign direct investment (FDI) in developing regions, 1980–2005. (From UNCTAD Foreign Direct Investment Database.)

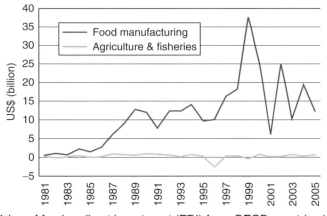

Figure 8. Value of foreign direct investment (FDI) from OECD countries in agriculture and fisheries and food manufacturing sectors, 1981–2005. (From OECD International Direct Investment Statistics.)

products. This is illustrated further by the level of cross-border mergers and acquisitions in certain agro-industrial sectors in developing regions (Table 2).

The increasing flows of FDI to developing countries serve not only to alleviate capital constraints on processes of industrialization (Reardon and Barrett, 2000) but also to facilitate the flow of new technologies and management practices, and to induce more efficient paths to organizational and institutional change. Investment flows by agro-industrial enterprises in industrialized countries can also be an effective mechanism to 'capture' more advanced technologies and management systems. At the same time, large inflows of capital from foreign enterprises can bring about rapid processes of concentration in agro-industrial sectors and, in due course, significant capital outflows in the form of expatriated profits. For example, the entry of Nestlé and Unilever into China, bringing their own proprietary standards for food safety and quality, induced domestic firms to implement equivalent standards and to adapt their managerial and marketing systems (Wei and Cacho, 2001; Reardon, 2007). In turn, leading domestic firms were able to increase their competitiveness in the domestic market, such that they were able to capture market share at the expense of the multinationals and weaker domestic enterprises, with the effect that overall market concentration increased. More generally, the entry of foreign competitors can have profound impacts, not only on the agro-processing sector itself, but also on the entire supply chain; the example of dairy processing in Brazil in Box 1 provides a good illustration.

While processes of transformation in the agro-industrial sector are well-established in many developing countries, these processes have more recently been taking place in the context of, and have been further induced by, the restructuring of food retail markets. Key here is the growth of the supermarket sector. While supermarkets have long operated in a number of developing countries, these were generally confined to large cities and focused on upper-middle or rich consumer segments. However, there is some evidence that a

Table 2. Distribution of cross-border mergers and acquisitions in developing regions by sector, 2004–2005 (US$ million). (From UNCTAD, 2006.)

Sector	Africa				South, East and South-east Asia				West Asia				Latin America			
	Sales		Purchases		Sales		Purchases		Sales		Purchases		Sales		Purchases	
	2004	2005	2004	2005	2004	2005	2004	2005	2004	2005	2004	2005	2004	2005	2004	2005
Total	4,595	10,509	2,718	15,505	24,193	45,132	19,319	35,349	575	14,134	1,280	18,221	21,840	22,532	11,977	10,179
Primary	3,994	908	1,680	249	421	469	819	4,312	383	111		45	1,333	814	8	881
Manufacturing	68	1,676	529	35	7,386	13,300	4,769	14,805	146	55	922	19	6,560	10,793	8,582	5,492
Food, beverages and tobacco	46	17		3	1,575	6,256	373	7,040					4,131	5,710	7,786	127
Wood and wood products		120	452		320	997	162	30								
Tertiary	533	7,925	509	15,221	16,385	31,363	13,730	16,222	46	13,968	357	18,157	13,947	10,926	3,322	3,806

> **Box 1.** Transformation in the dairy processing sector in Brazil.
> (From Reardon, 2007.)
>
> The Brazilian Government's decision to liberalize foreign direct investment (FDI) in 1990 brought about profound and rapid changes to the structure and conduct of the dairy products sector. The relaxation of restrictions on foreign investment facilitated the entry of MD Foods, Parmalat and Royal Numico, alongside new investments by Nestlé, which had a longer-term foothold in the country. In order to gain market share these new entrants engaged in aggressive price competition that had the effect of reducing retail milk prices by 40% over the period 1994–1997. At the same time, efforts were made to reduce costs throughout the raw milk supply chain. These efforts included enhanced efficiencies alongside the imposition of private standards to improve safety and quality in order to promote milk and dairy product consumption and reduce losses during processing, predominantly through lower bacterial counts. Within 5 years these standards had been rolled out across the supply chains of the major dairy processing companies, expelling thousands of small dairies that were unable to make the required investments in refrigerated storage vessels.

'supermarket revolution' has been under way in certain developing countries since the early to mid-1990s (Reardon *et al.*, 2007), although with significant variations across developing regions. The penetration of supermarkets is greatest in South America, South Africa and East Asia outside China. In these countries, the growth of supermarkets started in the early 1990s and their average share in food retail sales grew from around 10–20% in 1990 to 50–60% by the early 2000s (Reardon and Berdegué, 2002). In a 'second wave' of countries, including Mexico and much of South-east Asia and Central America, the rapid growth of supermarkets did not occur until the late 1990s and their current market share is in the region of 30–50%. However, in many other developing regions, for example, eastern and southern Africa and poorer countries in Asia and South and Central America, supermarkets are just emerging in major cities, while their operations tend to be confined to packaged groceries. Indeed, doubts have been raised about the pace of retail food market restructuring in these areas (see e.g. Humphrey, 2007), suggesting that supermarkets will remain the exception rather than the norm in the short to medium term at least.

The growth of the supermarket sector in developing countries has been induced by many of the same trends influencing the evolution of the agro-industrial sector, including changing demand patterns, liberalization of domestic and international food markets and FDI (Reardon, 2007). In turn, the transformation of the food retail sector is serving to 'amplify' these trends and induce changes in the structure and organization of agro-industrial enterprises and their downstream relations in supply chains. Thus, as supermarket procurement systems develop and evolve (Reardon *et al.*, 2007), there is demand for larger supply volumes and competitive advantages from the acquisition of enhanced skills in food safety and quality standards and supply chain management that tend to favour larger enterprises. Indeed, given that processed food products constitute 65% of supermarket food sales in developing countries, and semi-processed food products account for a further 20–25%, the development of the supermarket sector is dependent on appropriate responses by the food manufacturing sector, at least initially creating conditions of mutual dependency. However, as supermarkets come to command a greater and greater share of the retail market for food, and their distribution systems begin to spread beyond national boundaries, there is a definite shift of power such that supply chains for processed food products are increasingly buyer (i.e. retailer) driven.

The evolution of the agro-industrial sector along the lines described above induces changes in the skill sets required to compete. Modes of management that predominate in agro-industrial sectors have shifted from the simple coordination of product flows, management of processing operations and transfer of ownership, to the implementation of closely attuned systems of production, processing and distribution that increasingly span the length of the supply chain, not only nationally but internationally also. While such changes have been facilitated by technological advances, among other things, they are also driven by the needs and demands of dominant actors (Dolan et al., 1999; Dolan and Humphrey, 2000). Existing firms must acquire the related skills in order to flourish, or face being competed out of the market. In particular, there is increasingly a premium on management skills (Reardon and Barrett, 2000) across all of a firm's operations from raw material procurement, through processing operations to marketing, and including firm strategy and labour relations. While multinationals are generally able to import the required skills from their operations elsewhere, domestic firms must either develop these skills internally or rely on external service providers. Thus, accompanying processes of agro-industrialization is the emergence of providers of managerial and other business services. For example, in coastal Peru, firms provide labour supervision services over small cotton farmers in return for land collateral under arrangements that resemble share tenancy contracts (Escobal et al., 2000). In other cases non-governmental organizations provide such services, acting as an implicit subsidy supporting processes of agro-industrialization, for example, by acting as an 'honest broker' in establishing supply relations between agro-processing enterprises and primary producers.

Development of the agro-industrial sector and related changes in the structure and conduct of markets for food and non-food products can induce considerable changes in vertical relations along supply chains, most notably with primary producers. Thus, the predominance of spot markets with informal rules of conduct for many agricultural, fisheries and forestry products is giving way to more formal and longer-term relationships between agro-industrial enterprises and suppliers of raw materials. In particular, contracting is becoming a key form of governance in vertical supply chains, while forms of contracting are changing. Initially, agro-industrial enterprises tend to employ marketing contracts that are relatively informal in nature, or at least are not specified in a written and legally enforceable instrument. Over time, however, contracts tend to encompass elements of the entire production process and are increasingly codified in written standards and terms of supply. While such contracts act to enhance the security of raw material supplies for agro-industries, they also provide a mechanism for reducing and redistributing risk along the supply chain. In turn, they also induce changes in the nature of primary production, typically requiring the increased use of off-farm inputs. Thus, primary producers are increasingly integrated into the commercial supply chain for both output marketing and input supply, inducing the evolution of allied supply sectors (seed, fertilizer, etc.). At the same time, there are pressures towards the consolidation of primary production, whether through larger farms or collective action on the part of small farmers, raising the spectre that smallholders will be progressively excluded from supply chains as agro-industrialization proceeds.

Increasingly tied to contractual relations with raw material suppliers are quality and safety standards that not only stipulate end-product characteristics but also define elements of the production process (Henson and Reardon, 2005) as a means to ensure that required processing parameters are complied with and the requirements of downstream buyers are satisfied. As these standards are 'ratcheted up' over time, there are profound implications for the position of small farmers in the supply chain, in terms of both their ability to participate in supply chains to agro-industrial enterprises and the impact on rural incomes. As we will see below, the evidence in this regard is marked, providing instances of exclusion from supply chains, but also examples of small-scale producers adapting to the changes required and gaining through higher income as a result.

Despite the very visible and enticing trends towards industrialization of the agro-processing sector outlined above, in many developing countries (and especially low-income countries), informal agro-processing remains strong, and indeed continues to be the predominant locus of agro-processing enterprises (Sautier *et al.*, 2006). This situation reflects the fact that the entry costs to informal agro-processing tend to be low, in view of the fact that levels of capitalization are typically limited and traditional technologies command only low levels of human capital that can be acquired relatively easily through apprenticeships or other traditional means. Thus, informal agro-processing enterprises tend to be highly responsive to market demands within the low-income markets in which they tend to predominate, while innovative capacity tends to be greater than is often recognized. These characteristics, alongside the numerous capital, managerial and technological constraints faced by these businesses, are reflected in the fact that rates of mortality of informal enterprises are high (Mead, 1994; Mead and Liedholm, 1998).

The degree of structural and organizational transformation of the agro-industrial sector differs by country and even by regions of a country. In general terms, agro-industrialization has advanced most in countries that have achieved the greatest level of integration into global supply chains for higher-value food and/or non-food products, and/or where domestic high-value markets have evolved in response to economic, social and demographic change. In the latter case, this relates to broader processes of economic growth and development, and thus is typically found in countries with higher per capita incomes. However, even in very poor countries with low levels of overall economic development and where supply chains are predominantly traditional in nature, it is possible to find 'enclaves' where transformed and dynamic agro-industrial sectors exist. Kenya provides a good example; while the informal sector predominates in agro-processing, there is a relatively well-developed industrial dairy processing sector directed at domestic and regional markets and a number of globally competitive exporters of semi-processed vegetables.

Among poorer countries, including the least-developed countries of sub-Saharan Africa, agriculture still represents a large proportion of value added along agro-food supply chains. Traditional supply chains for agro-food products generally predominate and high-value domestic markets are in their infancy; an indicator of this is the low market penetration of supermarkets in most countries of sub-Saharan Africa. Usually, the formal agro-processing sector is small, and may even be stagnating, and there is little or no integration along the supply chain. There are, however, exceptions to this general rule (e.g. Zambia and Ghana), where high-value markets are more pronounced due to foreign investment in food processing and/or supermarkets, significant levels of remittances, burgeoning middle- and high-income groups, etc. Further, in a number of low-income countries supply chains have been established, predominantly to supply high-value markets in Europe (such as green beans in Kenya). However, such supply chains remain the exception rather than the rule, and operate within a 'sea' of fragmented and multilayered traditional markets that are 'ruled' by the informal sector.

In middle-income (and especially upper-middle-income) countries, agriculture typically accounts for a small proportion of value added along agro-food supply chains. Here the process of agro-industrialization is generally more pronounced and widespread. Urban high-value markets are typically well developed, sometimes with high levels of supermarket penetration (e.g. Brazil, Mexico, Malaysia, Egypt and Thailand). In many cases, there is also a vibrant agro-processing sector that has evolved in response to the growth of domestic market demand and/or competitive 'gaps' in international markets, for example, for processed foods (as in the case of canned tuna production in Thailand). A number of these countries are also significant exporters of higher-value agro-food products and are integrated into the associated global supply chains. However, supply chains to domestic and international high-value markets often coexist and operate independently of one another. Indeed, it is possible to observe a continuum in the level of transformation, with traditional agro-food sectors predominant in rural markets, while a whole spectrum of formal and informal sub-sectors supplying domestic markets exists in between.

Impacts of Processes of Agro-industrialization

Processes of agro-industrialization have widespread and profound impacts at both micro and macro levels (Figure 1). These include contributions to overall economic development, alongside changes in rates of poverty linked to the scale and distribution of changes in employment and per capita incomes among those whose livelihood is linked to the agro-food economy. These processes also encompass the quality, availability and price of food and non-food products, plus impacts on natural resources and the environment, and sociocultural implications, among others. Thus, we might reasonably expect there to be gainers and losers from processes of agro-industrialization, such that there are likely to be significant distributional consequences of the emergence of an agro-industrial sector. What is critical in this context is to recognize and promote the conditions under which agro-industrial enterprises can make a positive and significant contribution to overall processes of economic development and to the betterment of the lives of the poorest members of society, while minimizing any negative externalities and other impacts.

Given that agro-industrialization is but one of the multiple transformations occurring in agro-food systems in developing countries, as well as globally, attributing observed changes to the development of an industrial agro-processing sector per se is difficult. Further, these impacts reflect the ongoing dynamics of development processes, for example, the transformation of a predominantly informal to a predominantly formal economy, and observations at any point in time merely provide a 'snapshot' that is soon outdated.

The impacts of agro-industrialization on value addition, exports and employment are treated at length in Chapter 3 of this volume. We will instead call attention to the effects of agro-industries on two specific aspects: power relations along the supply chains and environmental consequences of agro-industrialization.

Impacts on power relations along the supply chain

One of the key concerns related to the processes of consolidation and globalization accompanying agro-industrialization is the impact on power relations along supply chains and the degree to which dominant players, including agro-processing firms and their purchasing agents, operate in a competitive environment or are able to exert 'unfair' power (Vorley and Fox, 2004). Indeed, some critics argue that agro-industrialization and related processes of concentration tend to develop in a mutually reinforcing cycle along agricultural and food supply chains (Lang, 2003). Thus, as markets for higher-value agro-food products develop, a limited number of dominant firms tend to emerge that quickly 'quash' their smaller competitors in the formal sector while commanding a progressively greater proportion of the market as the informal sector is 'squeezed out'. In some cases these dominant enterprises are specific to particular markets, while in others their influence cuts across many geographical regions and/or product sectors. There are related concerns about the distribution of rents along supply chains that are under the influence of such entities and the scope for primary producers (and especially small-scale farmers and fishers) to influence the conduct and performance of supply chains and enhance their share of value added through upgrading (Rabellotti and Schmitz, 1999; Kaplinsky, 2000; Humphrey and Schmitz, 2001, 2003).

In the context of many developing countries, the scope for concentration of markets to bring about 'abuse' of market power is exacerbated by the fact that regulatory controls tend to be weak. Thus, competition policy frameworks are typically underdeveloped, if present at all, such that there is no legal basis through which the power of dominant players can be 'reigned in'. Further, challenging business conditions (see below) tend to act as barriers to entry, especially for smaller firms that lack economic and political power to thwart bureaucratic delays in official and/or legal procedures, such as the enforcement of contracts.

At the same time, there is a need to consider the 'counterfactual' to these trends towards consolidation of agro-industrial sectors; that is, what would be the situation along supply chains for agro-food products were there to be no process of agro-industrialization? Further, the evolution of agro-industries needs to be viewed in the context of broader processes of transformation in global, national and rural economies (Reardon and Timmer, 2005). On the one hand, agro-industrial enterprises play a key role in the evolution of higher-value markets for agro-food products, which serves to stimulate demand for the products of primary producers. On the other, in the absence of agro-industries as a key driver of supply chains for higher-value agro-food products, the power vacuum might be filled by large multinational traders. The emerging supermarket sector may also have more command over food supply chains in the absence of large agro-industries to 'check' their market power.

Environmental impacts of agro-industrialization

While the impact of agro-industrialization on the environment has begun to receive attention in recent years, there is relatively little supporting empirical evidence on it. Broadly, agro-industrialization usually carries with it an expansion in scale at various levels of the value chain. Expanded scale can come about through consolidation of existing establishments, new arrangements with small producers or firms (e.g. out-grower schemes) or new entrants (either from within the country or multinationals). It is often the case that expanded scale of operation brings about the adoption of new technologies, organizational forms and management approaches, which can carry with them positive environmental effects. These positive effects can be offset, however, by the degradation of the natural resource base and the production of environmental externalities that prevailing control capacities are unable to manage. More broadly, it is important to recognize that the impacts of processes of agro-industrialization as a whole reflect interconnected processes of change at various levels of agro-industry from production through to distribution. Thus, Barrett et al. (2001) suggest that we should examine the environmental impacts of agro-industrialization through three distinct lenses: (1) direct effects on agriculture and upstream supply industries; (2) direct downstream effects on processing, distribution and related commercial activities in agro-industrial supply chains; and (3) indirect effects, such as income growth and other structural changes.

Turning first to the direct effects on agriculture and downstream supply industries, the increases and transformations in agricultural production that accompany agro-industrialization have profound implications for land usage. Increased production often comes about by bringing more marginal and potentially sensitive land into cultivation, thus leading to concerns about deforestation, desertification and loss of biodiversity (among others), and/or the impacts of intensification via the adoption of new technologies on the existing land base. Evidence suggests mixed environmental effects arising from land expansion and/or intensification (see e.g. Barbier, 2000; Lee and Barrett, 2000), with the multifaceted drivers of land use (e.g. other commercial uses and urbanization) making it difficult to predict the land-use impacts of agro-industrialization. For example, we might see very different outcomes in scenarios where broader processes of industrialization compete for land than in cases where agro-industrialization occurs in the context of relative plentiful supply of land.

Likewise, the environmental impacts of agro-industrialization vis-à-vis other agricultural inputs can be both positive and negative. Agro-industrialization can induce changes in the type and/or level of usage of agrochemicals (see e.g. Dasgupta *et al.*, 2001), which may be negative if overall usage increases, but positive if more advanced and safer active ingredients are used, with the overall impacts dependent on the relative magnitude of these two effects. At the same time, adoption of technologies leading to capital-for-labour substitution can reduce employment and generate negative income effects for some, which may necessitate the use of production practices by food-insecure farmers that are far from sustainable. However, greater vertical integration via agro-industrialization can also mean that signals reach producers that otherwise would not be transmitted. For instance, multinationals that procure agricultural product from developing countries may be sensitive to consumer concerns over environmental impacts, and thus promote the adoption of environmentally friendly production practices, often with associated price premiums in niche organic markets (World Bank, 2005).

Processes of agro-industrialization can have critical effects on the availability and quality of water in developing countries. While increases in production may imply an increase in demand for water, especially if they are associated with irrigated systems of production, Barrett *et al.* (2001) suggest that agro-industrialization often brings about substitution of less water-intensive and higher-value crops for water-intensive cereals, thus presenting scope for water conservation. Conversely, some of the major higher-value food exports from developing countries, for example, fresh fruits and vegetables, require large volumes of water in their production. This fact has brought about accusations that, in effect, these countries are exporting 'virtual water' (Orr and Chapagain, 2007). Water pollution can become an issue with respect to pesticide use and (perhaps more importantly) livestock production. Many developing countries lack the institutions needed to properly develop and implement environmental governance systems to keep such pollution in check. On the other hand, livestock production plays an important role in converting organic matter into green fertilizer, the use of which leads to reduced application of agro-fertilizers and can lead to soil nutrient improvements and mulching that reduces water losses. Clearly, the net effects of agro-industrialization on water use and quality are both complex and uncertain, and certainly context-specific.

Broadly, little is known about the environmental effects of direct downstream elements of agro-industrialization, although three clear areas of concern exist: (1) air and water pollution associated with processing and distribution; (2) levels and nature of post-farm-gate solid waste; and (3) energy use. While agro-processing is typically one of the most polluting industries in developing countries (Barrett *et al.*, 2001), it is possible that processes of agro-industrialization can reduce certain aspects of its 'environmental load'. This might occur through the *de novo* entry of firms with cleaner processing technologies and/or from technological upgrading by existing firms. Further, economies of scale are important in the agro-processing sector, with larger scale of operation often being associated with increased industrial concentration. Both greater scale and concentration can make it easier to regulate the environmental impacts of agro-processing operations, with larger firms being easier to monitor and to take enforcement actions against (Lanjouw, 1997; Jayaraman and Lanjouw, 2000). The trade-off, however, is that scale of operation also creates incentive for firms to lobby for less strict regulatory controls or, conversely, for larger (often multinational) firms to lobby for stricter regulations that exclude smaller (often domestic) firms.

There are a number of potentially detrimental environmental effects associated with waste from the agro-processing sector, both in the pre-industrial and industrial phases. On the one hand, there are the waste materials from processing operations, some of which may be utilizable as by-products and others that require disposal. Barrett *et al.* (2001) note that industrialization of agro-processing not only creates a new (and often enhanced) stream of waste, but also distributes this waste away from the point of raw material production to other locations. The scope for the utilization of waste changes over time and can be context-specific. For example, new technologies and markets can evolve that enable waste materials to be used productively. The recent development of biofuel technologies and new ways in which materials can be transformed for use as animal feed provide good examples. On the other hand, income growth and the parallel transformation of agro-industries tend to increase consumption of packaged food products. Many of these packaging materials are imported into developing countries and ultimately have to be disposed of in some way. In many urban areas of developing countries, food packaging materials have become a major environmental issue, for example, through blocking water courses and sewage systems.

Processes of agro-industrialization have profound impacts on both the level of energy uses and the relative importance of different sources of energy, with often mixed environmental impacts. Broadly, agro-industrialization tends to lead to substitution of capital for labour, and thus increases the overall demand for energy. However, net energy savings can occur if commercial-scale processing, with more energy-efficient technologies, replaces less energy-efficient technologies that would otherwise occur on a smaller (and often informal) scale. While agro-industrialization will tend to increase the demand for transport and induce greater use of fuel, especially where it implies geographical specialization, the adoption of more fuel-efficient transport modes could have offsetting effects. Moreover, given that industrialized processing facilities often locate near raw material sources, potential energy savings could be realized by shipping finished goods rather than raw materials.

While the direct and multifaceted impacts of the industrialization of agro-processing are uncertain, potentially the more general and indirect effects are more critical. Notably, we might reasonably expect the environmental effects of agro-industrialization to follow the predications of the environmental Kuznets curve: as developing country incomes grow, environmental degradation will first increase, reach a peak and then begin to decline, reflecting the adoption of 'cleaner' technologies to abate or remedy negative environmental impacts. Barrett *et al.* (2001) suggest that the balance of current evidence points to increased environmental degradation alongside processes of economic development and agro-industrialization, with little indication of a shift to latter parts of the Kuznets curve. Thus, while the scale effects associated with processes of agro-industrialization play a role in increased environmental degradation, it may well be that the adoption of cleaner processing technologies and production practices has not progressed to the extent that the detrimental environmental effects are offset.

Challenges Faced with Ongoing Processes of Agro-industrialization

The foregoing discussions and some later chapters highlight some of the impacts that agro-industrialization can have on developing countries, outlining the positive contributions that agro-industrial enterprises can play, but also some of the detrimental effects. The challenge for developing countries in steering the development of agro-industrial sectors is to 'clear the way' for private investments that will bring about rapid evolution of agro-food systems, while laying down conditions that will enable the potential negative consequences to be minimized. This suggests that these challenges lie not only within the narrow frame of the agro-food sector, and agro-processing in particular, but in the broader economic and institutional environment in which industrialization takes place. Below we consider a number of the key challenges in turn.

Global positioning

The preceding discussion highlighted how 'the world is changing'. Processes of agro-industrialization in developing countries are taking place within the context of agro-food systems that are increasingly globalized and themselves undergoing organizational and institutional restructuring. While prevailing trends can present new opportunities for developing countries in the form of value addition, they also pose challenges. Thus, the bases on which developing country agro-industrial enterprises generally competed in the past are increasingly becoming obsolete, while these enterprises are increasingly facing competition from their global counterparts. At the same time, opportunities for value addition in this 'new world' require new approaches to competitiveness. Consequently, we are likely to see very different forms of agro-industrial enterprise evolve in the future, while existing enterprises that are unable to compete in this 'new world' will shrivel away.

Perhaps the most critical challenge for developing countries is to identify where their agro-industrial sectors fit into this new 'global order'. This implies that the status quo will not suffice. Rather, agro-industries must be driven primarily by consumer demands, as they filter down through downstream actors (e.g. supermarket chains and processing enterprises). In many cases this implies a shift of thinking that is even more radical than that required when moving away from the statist model of development that prevailed until the mid-1980s (Jaffee et al., 2003). Indeed, in many cases, agro-industries, as well as the wider agro-food system, will need an entire reorganization from 'supply-push' to 'demand-pull', recognizing the primacy of consumers and dominant downstream buyers. Undoubtedly this poses huge challenges, and the associated processes of organizational and institutional restructuring are immense; the more so given the resource and infrastructural constraints faced by governments and agro-industrial enterprises in many developing countries.

It is imperative that developing countries recognize the new realities of the global agro-food economy and identify how they can position existing and/or new agro-industries within this reality. The challenges presented by the contemporary agro-food economy are extremely daunting, especially where costly processes of restructuring and upgrading are required in order to compete in both evolving and emerging value chains. At the same time, however, there is some evidence that the challenges that agro-enterprises are facing, for example, in order to comply with stricter food safety and quality standards in international markets, can act as fundamental catalysts of change and strategic repositioning (World Bank, 2005; Henson and Jaffee, 2008). Indeed, it is argued that developing countries are more likely to succeed where there is little or no 'latitude for poor performance'; that is, where requirements are laid out specifically and where there is little or no tolerance for deviation from these standards (Crammer, 1999).

More generally, we can look at the experiences of a number of developing countries that have been successful in accessing higher-value markets. Certain countries have managed to add value to traditional agro-food exports through agro-processing, for example, Côte d'Ivoire (fisheries and wood), Senegal (fisheries) and Ghana (wood) (Crammer, 1999). Other countries have diversified diagonally by shifting from traditional primary exports to the processing of other products, for example, Equatorial Guinea (cocoa to sawn wood and veneer sheets) and Kenya (from coffee and tea to horticultural and fisheries products). Much can be learned from these experiences, and also comparable efforts that have failed.

Infrastructural constraints

Preconditions for the development of agro-industries are the necessary transportation, information and communication technologies (ICT) and access to reliable supplies of key utilities, notably electricity and water. In turn, the infrastructural constraints under which the agro-industrial sector operates influence the cost and reliability of the physical movement of raw materials and end products, efficiency of processing operations, responsiveness to customer demands, etc. Indeed, alongside prevailing macroeconomic and business conditions, the level, quality and reliability of infrastructure have been shown to be a critical determinant of export competitiveness for processed agro-food products (Crammer, 1999). Where infrastructural constraints are particularly acute the additional complexities of processing operations may outweigh the benefits of diversification away from exports of primary commodities and towards value addition (Love, 1983).

In many developing countries, and especially low-income countries, infrastructure tends to be weak. Inherently, this puts agro-industrial enterprises at a competitive disadvantage to their industrialized country competitors, while also distorting the competitiveness of developing countries relative to one another according to the quality of their basic infrastructure. Thus, agro-industrial enterprises may face unreliable and costly transportation systems that prevent access to potentially lucrative markets. Under such conditions we might find potentially competitive agro-processing firms that are unable to access key markets because of the weakness of transportation systems. Likewise, unreliable and costly supplies of utilities can prevent agro-processing companies from operating at or near full capacity utilization.

The existence of weak infrastructure can influence the rate of transition of the agro-processing sector from informality to formality, and the evolution of the sector's structure over time. Thus, the adoption of more advanced technologies, which is one dimension along which the formal agro-processing sector competes with often lower-cost informal enterprises, is dependent on reliable access to electricity and water. Without access to essential inputs and utilities the agro-processing sector can be caught in an informality 'trap'. Further, weak infrastructure tends to favour larger enterprises that have access to the capital to install their own facilities for generating electricity and providing potable water, and operate at capacity levels to spread these costs over a large volume of output. In the longer term, as processes of agro-industrialization proceed, this can steer the structure of the sector towards higher levels of concentration.

In addition to basic infrastructure such as roads, electricity, the Internet and telephones, more specific infrastructural needs of the agro-industries sector continue to develop. Examples include access to laboratory testing and certification services, providers of repair services and new product development facilities. In many developing countries such infrastructure is weak, such that compliance with even basic food safety and quality standards can be problematic. Undoubtedly, this acts to impede competitiveness relative to agro-industrial enterprises that have better access to such services or, at the minimum, increases costs as firms are forced to make use of service providers in neighbouring (or even distant) countries. It can also force sectors into reactive modes of upgrading capacity when 'problems strike' rather than more proactive approaches that maximize market competitiveness.

Illustrating the critical role of both general and agro-processing-specific infrastructure to the evolution of the agro-industrial sector, Jaffee and Morton (1995) highlight how the failure of many large-scale parastatal processing industries in sub-Saharan Africa can be explained, at least in part, by their inability to access essential support services (e.g. repair of machinery), and by their unreliable access to utilities, etc. Indeed, many of these enterprises operated at well below their level of installed capacity or had ceased operations altogether, often due to factors that were outside of their control. Arguably, many of these enterprises would have drifted into bankruptcy if they had not been provided with protection from governments and/or financial support from donors.

Access to physical and human capital

With the progressive shift from the informal to formal sectors, and as agro-processing enterprises attempt to add value and compete with their industrialized country counterparts, access to the required physical and human capital becomes more critical. Indeed, processes of agro-industrialization are associated with, and at the same time themselves induce, technological changes along the supply chain, for example, through improved crops and livestock, new forms of processing and enhanced distribution systems. Such changes are critical in order to achieve improvements in efficiency, meet the evolving demands of buyers and consumers and enhance storability and transportability. Simultaneously, the very nature of these technologies is changing, as is well illustrated by advances in ICT and the increasing use of biotechnology.

Agro-industries in developing countries often face significant problems in gaining access to the technologies and skills they require in order to evolve and compete in the contemporary agro-food economy, either because these are not available domestically or because they are costly. In many cases these technologies are imported, although import taxation regimes, access to foreign exchange and the exchange rate can act as significant impediments. This reflects the fact that research and development expenditure in many developing countries is low. Alternatively, the transfer of physical and human capital can occur internationally, through linkages with multinational corporations, technical assistance provided by bilateral or multilateral donors, etc. Indeed, there is mounting evidence that firms in developing countries can accrue critical capacities through their interactions with international buyers (Schmitz and Knorringa, 2000). Critical here is that technologies and skills are 'appropriate' to the specific context of the developing country concerned and to the position of a particular industry and/or enterprise in domestic or global markets. Thus, for example, a highly sophisticated and costly technology may not be appropriate for a firm or industry that is pursuing a cost leadership strategy. Further, where technologies are highly product-specific, high levels of asset specificity can make such investments risky and deter potential investors.

While low levels of capital intensity that result from impediments to accessing technologies, alongside labour endowments, are often posited as reasons for the lack of international competitiveness of developing countries in manufactured product exports (Crammer, 1999), there is mounting evidence that the level of skills per worker (human capital) and land endowments are more critical for processed primary agricultural products (Wood and Berge, 1997; Wood and Owens, 1997). This suggests that knowledge-based competitive assets (as embodied in the labour force) are critical to processes of agro-industrialization in developing countries. This, in turn, argues for a need for investments in general and skill-specific education. Indeed, this would seem logical; while access to technology is critical, no technology can be employed unless there is the trained workforce to operate it.

Macroeconomic and policy environment

Alongside the specific challenges faced by the agro-industrial sector, all enterprises have to operate in the more general macroeconomic, legal and policy environment. Under conditions of macroeconomic instability, capital investments with significant sunk costs, as with technologies that are highly product-specific, are deterred by the inherently greater risk. This situation can then be further exacerbated with exchange rate misalignments that enhance uncertainty. Where technologies and critical inputs have to be imported, foreign exchange shortages linked to import licensing and foreign exchange allocation systems can delay investments or impose additional costs on entrepreneurs, such as in the form of 'lobbying' for foreign exchange.

In many developing countries government policies have been slow to adapt to the 'new reality' of private enterprise. For example, although most countries have undergone some process of structural adjustment, liberalization of controls on investment and trade is uneven. Further, licensing of businesses remains common. Variations in the interpretation and implementation of these policies serve to create uncertainty and add costs (e.g. in the form of bribes) and delays to business. Such conditions are far from conducive to private investment and can act to deter foreign firms from investing at all, while discouraging more risky (but also higher yielding) investments, such as are often associated with more advanced technologies. These aspects are explored in more detail in Chapter 5 of this volume.

Business conditions

Although related to the broader policy environment, general business conditions are a specific factor determining the rate and trajectory of the development of the agro-industrial sector. Thus, lengthy and costly procedures for the legal registration of a business and enforcement of contracts can act as a barrier to the transition of enterprises from the informal to the formal sector (De Soto, 1989) and provide disincentives for private investment in new enterprises. This particularly applies where agro-industries are capital-intensive and require large initial investments; under such conditions the timely start-up of a business can be critical for commercial viability.

Table 3 provides selected measures of the business conditions in developing and industrialized countries. It is immediately apparent that, while there are significant inter-country differences within any one country income group, the procedures and time required to start up a business and time required to enforce a contract are significantly greater in developing countries, especially in low-income countries. While efforts have been made to establish regulatory frameworks that are more conducive to business in many developing countries, bureaucratic inertia and a lack of effort to stamp out corruption have served to maintain conditions that challenge private enterprise.

A second element of the conditions required for business is a well-developed and reliable financial and other business service sector. For example, where the banking sector is weak, resource mobilization is hampered by low credit repayment rates, which in turn become reflected in higher interest rates, high transaction costs and/or political interference. Problems in gaining access to fixed and/or working capital, including trade credit, are most important for smaller businesses that tend to lack the required financial resources internally, with the result that unreliable banking systems can deter the transition from the informal to formal sector.

Table 3. Indicators of business conditions by country income group, 2006. (From World Bank World Development Indicators.)

Country/country group	Start-up procedures to register a business (number)	Time required to start a business (days)	Time required to enforce a contract (days)
All developing countries	10.2	54.0	425.6
Low-income countries	10.5	61.5	428.5
Zambia	6.0	35.0	274.0
DRC	13.0	155.0	909.0
Lower-middle-income countries	10.4	52.2	432.9
Tunisia	9.0	14.0	27.0
Angola	14.0	146.0	1011.0
Upper-middle-income countries	9.0	42.2	405.0
Turkey	8.0	9.0	330.0
Venezuela	13.0	116.0	445.0
High-income countries	7.1	23.2	277.5
Australia	2.0	2.0	157.0
Slovenia	9.0	60.0	913.0

Global trade regimes

While the expansion of domestic demand for higher-value agro-food products is likely to be an increasing catalyst of processes of agro-industrialization, the ability of developing countries to supply global markets will remain a critical issue (Diaz-Bonilla and Reca, 2000). Indeed, it has long been held that a viable strategy for developing countries towards industrialization is the processing of primary commodities, which is predominantly the trade in which they are already engaged (Crammer, 1999). While we have discussed a number of challenges faced by developing countries in supplying export markets, especially in view of the rapid evolution of the global agro-food economy, traditional trade restrictions (e.g. in the form of tariffs and quantitative restrictions) remain a problem. Indeed, tariff escalation according to the level of processing remains a reality in many industrialized countries, acting to thwart the ambitions of developing countries to move up the value chain, as they are being advised by those very same industrialized countries to do. At the same time, non-tariff measures, including food safety and quality standards, are creating new challenges, especially for developing countries that lack the critical food safety and quality management infrastructure (World Bank, 2005; Henson, 2007a), which may prevent them from being able to exploit preferential market access as low-income countries. Ongoing global trade negotiations clearly have a role to play in addressing this issue.

A secondary impediment to developing countries attempting to access industrialized (and also developing) country markets for agro-industrial products is the considerable advantage enjoyed by firms that already have a market presence, for example, through information networks and market linkages. Indeed, developing country agro-industrial enterprises can struggle to integrate themselves into increasingly sophisticated and integrated global supply chains. This can mean that considerable time is required for developing country agro-industries to exploit trade opportunities. It is perhaps not surprising, therefore, that there has been little change in the countries of sub-Saharan Africa that dominate exports of non-traditional agricultural and food products, reflecting the fact that there have been few appreciable new entrants (Henson, 2007b). We observe similar patterns in the trade of higher-value agricultural and food products in other parts of the world, suggesting very significant first entrant advantages.

Strategic thinking

In recent history, agro-industrialization has gone through two broad stages of development in a developing country context. Prior to the era of structural adjustment, the public sector played a dominant role in steering the establishment of large-scale and often publicly owned enterprises in the pursuit of rather dubious objectives, including driving broader processes of industrialization. While many of these efforts are recognized to have failed, the subsequent phase has arguably been little better. Thus, in the pursuit of private investment, many countries have left the agro-industrial sector to evolve in a *laissez-faire* manner, with little or no strategic direction at the sectoral or sub-sectoral level. Indeed, the liberal agenda pursued by many developing countries has caused uncertainty and confusion over the legitimate role of government.

Too often government has taken a 'back seat', by simply observing the evolution of agro-industries without defining and steering the sector towards strategic objectives. Indeed, there is a need for developing countries to identify the most appropriate path for processes of agro-industrialization in their particular country context. They should identify areas where they can or cannot compete in domestic and global markets and also establish how the growth of agro-industries can contribute to economic development through employment creation, reduction in poverty, reduction in market prices, enhancement of food safety and quality, environmental protection, etc. Without the adoption of a strategic approach to agro-industrialization, any contribution to these development objectives is largely coincidental. At the same time, there is fear, and indeed some evidence, that agro-industries are becoming highly concentrated, crowding out undercapitalized domestic firms and small producers, substituting domestic equipment for imports and enriching urban elites at the expense of the rural poor (Jaffee *et al.*, 2003).

The challenge for developing countries is to establish effective working relations between the public and private sectors in order to define a path for the development of the agro-industrial sector that does not stifle private incentives, yet brings about broad-based and sustainable growth that creates wealth and enhanced human well-being. This suggests that government has a legitimate role to play, not in directing private sector investments, but instead in creating conditions that are conducive to private investment and innovation and steering the development of the sector, in broad terms, in the 'right direction'. In turn, this requires that fruitful relations are developed between government and private enterprises, on a collective as well as an individual enterprise level, based on trust and mutual understanding.

The challenges highlighted above suggest that a national strategy for agro-industrial development is likely to be wide-ranging. However, some of the critical issues such a strategy might cover are as follows:

- Work with agro-industries at the sub-sector and sector levels to define plans to enhance competitiveness in domestic and global markets.
- Work with large agro-industries to assist small firms and producers to meet their requirements.
- Work to eliminate institutional barriers to entry which inhibit entrepreneurial dynamism.
- Ensure effective competition between enterprises in the agro-industrial sector so as to ensure choice for primary producers and consumers and 'fair' prices.
- Work towards the enhancement of general and sector-specific infrastructure, working with the private sector where appropriate.
- Lay down a regulatory framework that facilitates business investment, promotes competition between agro-industrial enterprises and ensures 'fair' treatment of consumers and primary producers.

- Make strategic investment in research and development that, rather than being broad-based, are directed at identified areas of competitive advantage.
- Negotiate with international trading partners for market access and technical assistance directed at competing in markets where a competitive advantage has been identified.

In developing such strategies there is no need for developing countries to each start from scratch and to 'reinvent the wheel'. Developing countries, and the sectors and firms therein, should be encouraged to share experiences. Bilateral and multilateral donors and development organizations can play a role in facilitating experience sharing and in supporting processes of technology transfer, while FDI and the trade relations that firms in developing countries have with international buyers will be of increasing importance to capacity development at the enterprise level.

Conclusions

This chapter has sought to outline the nature of the process of agro-industrialization in developing countries and the challenges being faced, most notably as the global agro-food economy changes in response to a series of pervasive mega trends. It is evident that the future trajectory of agro-industries will be quite different to the past and, while developing countries are making efforts to establish nascent agro-industries, the world around them is changing rapidly. Certainly, the basis of competitiveness of agro-industries in developing countries in the future is likely to be somewhat different to the past, requiring the transition of existing enterprises and the adoption of new policies and approaches to the establishment of enterprises in the future. At the same time, this 'new world' is creating significant opportunities for developing countries that are able to respond appropriately.

It is evident that agro-industries can (and do) play a fundamental role in overall processes of industrialization and economic development, although there are significant (and not always positive) micro impacts and externalities. For example, while agro-industries present new opportunities for more secure and better paid employment, the transition from the informal to the formal sector inevitably involves dramatic changes in the structure of value chains and associated vertical and horizontal power relations. Thus, agro-industries can evolve in a manner that acts to exclude smaller formal sector enterprises and small primary producers, having detrimental structural and livelihood impacts. Likewise, while agro-industrialization can bring about environmental benefits, there is also significant scope for negative environmental externalities. This highlights the need for processes of agro-industrialization to follow an 'appropriate path', guided by appropriate policies and strategies.

The overall picture painted by this chapter is that the development of agro-industries in the contemporary context provides opportunities for significant gains to developing countries, although exploiting the opportunities in a manner that these gains exceed the losses is far from certain. Certainly, there is a political case for promoting agro-industrialization in developing countries, although at the same time the development of agro-industries must follow a clearly defined path that avoids the multiple pitfalls that inevitably go with such fundamental processes of economic and social change. This suggests a gradual approach rather than a 'great leap forward' in promoting the agro-industrial sector (Jaffee and Morton, 1995). In turn, strategies are needed nationally and internationally to ensure that developing countries have access to the knowledge and technologies necessary to shift their agro-food sectors from the informal to the formal sectors. Experiences need to be shared across developing countries so that the 'same mistakes are not repeated time after time'. The international community clearly has a role to play here, facilitating cross-border collaboration and access to markets, alongside the efforts of national governments to lay down conditions that are conducive to private enterprise, while establishing regulatory frameworks and policies that guide the evolution of the sector along the 'right path'.

The future trajectory of agro-industries in developing countries is uncertain. Yet there are plentiful opportunities and many benefits to be had from agro-industrialization. However, at the same time, if processes of agro-industrialization are not guided appropriately, the negative effects, both short term through exclusion of small farmers and informal enterprises and long term through vertical and horizontal concentrations of supply chains and environmental externalities, could be considerable. There is a political imperative to plot a positive way forward, looking at positive and negative experiences and successes and failures to date, and setting out a strategic direction that serves the needs of developing countries and the consumers they serve worldwide.

References

Allaire, G. and Boyer, R. (1995). *Le Grande Transformation*. INRA, Paris.

Barbier, E.B. (2000). Links between Economic Liberalization and Rural Resource Degradation in the Developing Regions. *Agricultural Economics*, 23 (3), 299–310.

Barrett, C., Barbier, E. and Reardon, T. (2001). Agro-industrialization, Globalization and International Development: The Environmental Implications. *Environment and Development Economics*, 6, 419–433.

Busch, L. (2000). The Moral Economy of Grades and Standards. *Journal of Rural Studies*, 16, 273–283.

Busch, L. and Bain, C. (2004). New! Improved? The Transformation of the Global Agrifood System. *Rural Sociology*, 69 (3), 321–346.

Busch, L. and Bingen, J. (2006). Introduction: A New World of Standards. In: Bingen, J. and Busch, L. (eds) *Agricultural Standards: The Shape of the Global Food and Fiber System*. Springer, Dordrecht.

Cook, M. and Chaddad, F. (2000). Agro-industrialization of the Global Agrifood Economy: Bridging Development Economics and Agribusiness Research. *Agricultural Economics*, 23 (3), 207–218.

Crammer, C. (1999). Can Africa Industrialize by Processing Primary Commodities? The Case of Mozambican Cashew Nuts. *World Development*, 27 (7), 1247–1266.

Cranfield, J., Hertel, T.W., Eales, J.S. and Preckel, P.V. (1998). Changes in the Structure of Global Food Demand. *American Journal of Agricultural Economics*, 80 (5), 1042–1050.

Dasgupta, S., Mamingi, N. and Meisner, C. (2001). Pesticide Use in Brazil in the Era of Agro-industrialization and Globalization.

Environment and Development Economics, *6* (4), 459–482.

De Soto, H. (1989). *The Other Path: The Invisible Revolution in the Third World*. HarperCollins, New York

Diaz-Bonilla, E. and Reca, L. (2000). Trade and Agroindustrialization in Developing Countries: Trends and Policy Impacts. *Agricultural Economics*, *23* (3), 219–229.

Dolan, C. and Humphrey, J. (2000). Governance and Trade in Fresh Vegetables: The Impact of UK Supermarkets on the African Horticulture Industry. *Journal of Development Studies*, *37* (2), 147–176.

Dolan, C., Humphrey, J. and Harris-Pascal, C. (1999). *Horticulture Commodity Chains: The Impact of the UK Market on the African Fresh Vegetable Industry*. IDS Working Paper 96. Institute for Development Studies, Brighton.

Escobal, J., Agreda, V. and Reardon, T. (2000). Endogenous Institutional Innovation and Agroindustrialization on the Peruvian Coast. *Agricultural Economics*, *23* (3), 267–277.

FAO (1997). *The State of Food and Agriculture 1997*. FAO Agriculture Series No. 30, Rome.

Fold, N. and Pritchard, B. (2005). Introduction. In: Fold, N. and Pritchard, B. (eds) *Cross-Continental Food Chains*. Routledge, London.

Gereffi, G. (1999). International Trade and Industrial Upgrading in the Apparel Commodity Chain. *Journal of International Economics*, *48* (1), 37–70.

Gereffi, G., Humphrey, J. and Sturgeon, T. (2003). The Governance of Global Value Chains. *Review of International Political Economy*, *12* (1), 78–104.

Henson, S. (2007a). The Role of Public and Private Standards in Regulating International Food Markets. *Journal of International Agricultural Trade and Development*, *4* (1), 52–66.

Henson, S. (2007b). *New Markets and Their Supporting Institutions: Opportunities and Constraints for Demand Growth*. RIMISP, Santiago.

Henson, S.J. and Jaffee, S. (2008). Understanding Developing Country Strategic Responses to the Enhancement of Food Safety Standards. *The World Economy*, *31* (1), 1–15.

Henson, S.J. and Reardon, T. (2005). Private Agrifood Standards: Implications for Food Policy and the Agrifood System. *Food Policy*, *30* (3), 241–253.

Humphrey, J. (2007). The Supermarket Revolution in Developing Countries: Tidal Wave or Tough Competitive Struggle? *Journal of Economic Geography*, *7* (4), 433–450.

Humphrey, J. and Schmitz, H. (2001) Governance in Global Value Chains. *IDS Bulletin*, *32* (3), 19–29.

Humphrey, J. and Schmitz, H. (2003). Governance in Global Value Chains. In: Schmitz, H. (ed.) *Local Enterprises in the Global Economy: Issues of Governance and Upgrading*. Edward Elgar, Cheltenham.

ISO (2005). *The ISO Survey of ISO 9000 and ISO 14000 Certificates 2005*. International Organization for Standardization, Geneva.

Jaffee, S. and Morton, J. (1995). Private Sector High-Value Food Processing and Marketing: A Synthesis of African Experiences. In: Jaffee, S. and Morton, J. (eds) *Comparative Experiences of an Emergent Private Sector*. World Bank, Washington, DC.

Jaffee, S., Kopicki, R., Labaste, P. and Christie, I. (2003). *Modernizing Africa's Agro-Food Systems: Analytical Framework and Implications for Operations*. Africa Region Working Papers Series No. 44. World Bank, Washington, DC.

Jayaraman, R. and Lanjouw, P. (2000). *Small-scale Industry, Environmental Regulation and Poverty: The Case of Brazil*. World Bank, Washington, DC.

Kaplinsky, R. (2000). Globalization and Unequalization: What can be Learned from Value Chain Analysis. *Journal of Development Studies*, *32* (7), 117–146.

Lang, T. (2003). Food Industrialization and Food Power: Implications for Food Governance. *Development Policy Review*, *21* (5–6), 555–568.

Lanjouw, P. (1997). *Small-scale Industry, Poverty and the Environment: A Case Study of Ecuador*. World Bank, Washington, DC.

Lee, D.R. and Barrett, C.B. (eds) (2000). *Tradeoffs or Synergies? Agricultural*

Intensification, Economic Development and the Environment in Developing Countries. CAB International, Wallingford, UK.

Love, J. (1983). Concentration, Diversification and Earnings Instability: Some Evidence on Developing Countries' Exports of Manufactures and Primary Products. *World Development, 11* (9), 787–793.

Mead, D.C. (1994). The Contribution of Small Enterprises to Employment Growth in Southern and Eastern Africa. *World Development, 22* (12), 1881–1894.

Mead, D.C. and Liedholm, C. (1998). The Dynamics of Micro and Small Enterprises in Developing Countries. *World Development, 26* (1), 61–74.

Nadvi, K. and Waltring, F. (2003). Making Sense of Global Standards. In: Schmitz, H. (ed.) *Local Enterprises in the Global Economy: Issues of Governance and Upgrading.* Edward Elgar, Cheltenham.

OECD (2004). *Private Standards and the Shaping of the Agrifood System.* OECD, Paris.

Orr, S. and Chapagain, A. (2007). *Virtual Water Trade: A Case Study of Green Beans and Flowers from Africa.* International Institute for Environment and Development, London.

Pingali, P. and Khwaja, Y. (2004). *Globalization of Indian Diets and the Transformation of Food Supply Systems.* ESA Working Paper 04-05. Food and Agriculture Organization, Rome.

Ponte, S. and Gibbon, P. (2005). Quality Standards, Conventions and the Governance of Global Value Chains. *Economy and Society, 34* (1), 1–31.

Rabellotti, R. and Schmitz, H. (1999). The Internal Heterogeneity of Industrial Districts in Italy, Brazil and Mexico. *Regional Studies, 33* (2), 97–108.

Raikes, P., Jensen, M. and Ponte, S. (2000). Global Commodity Chain Analysis and the French Filière Approach: Comparison and Critique. *Economy and Society, 29* (3), 390–417.

Reardon, T. (2007). Global Food Industry Consolidation and Rural Agroindustrialization in Developing Economies. In: Haggblade, S., Hazell, P. and Reardon, T. (eds) *Transforming the Rural Nonfarm Economy: Opportunities and Threats in the Developing World.* IFPRI, Washington, DC.

Reardon, T. and Barrett, C. (2000). Agroindustrialization, Globalization, and International Development: An Overview of Issues, Patterns, and Determinants. *Agricultural Economics, 23* (3), 195–205.

Reardon, T. and Berdegué, J. (2002). The Rapid Rise of Supermarkets in Latin America: Challenges and Opportunities for Development. *Development Policy Review, 20* (4), 317–334.

Reardon, T. and Timmer, P. (2005). Transformation of Markets for Agricultural Output in Developing Countries Since 1950: How Has Thinking Changed? In: Evenson, R., Pingali, P. and Schultz, T.P. (eds) *Handbook of Agricultural Economics: Agricultural Development, Farmers, Farm Production and Farm Markets.* North-Holland, New York.

Reardon, T., Timmer, P., Barrett, C. and Berdegué, J. (2003). The Rise of Supermarkets in Africa, Asia and Latin America. *American Journal of Agricultural Economics, 85* (5), 1140–1146.

Reardon, T., Henson, S. and Berdegué, J. (2007). 'Proactive Fast-Tracking' Diffusion of Supermarkets in Developing Countries: Implications for Market Institutions and Trade. *Journal of Economic Geography, 7* (4), 399–431.

Regmi, A. and Gehlar, M. (2005). Processed Food Trade Pressured by Evolving Global Food Supply Chains. *Amber Waves,* February 2005, 1–10.

Sautier, D., Vermeulen, H., Fok, M. and Bienabe, E. (2006). *Case Studies of Agri-Processing and Contract Agriculture in Africa.* RIMISP, Santiago.

Schmitz, H. and Knorringa, P. (2000). Learning from Global Buyers. *Journal of Development Studies, 37* (2), 177–205.

Swinnen, J. and Maertens, M. (2007). Globalization, Privatization, and Vertical Coordination in Commodity Value Chains. *Agricultural Economics, 27* (s1), 89–102.

UNCTAD (2006). *World Investment Report 2006.* United Nations, New York.

Viciani, F., Stamoulis, K. and Zezza, A. (2001). Summary of Results of the Survey. In: Stamoulis, K. (ed.) *Food, Agriculture and Rural Development: Current and Emerging Issues for Economic Analysis and Policy Research*. Food and Agriculture Organization, Rome.

Vorley, W. and Fox, T. (2004). *Concentration in Food Supply and Retail Chains*. Paper prepared for the UK Department for International Development. International Institute for Environment and Development, London.

Wei, A. and Cacho, J. (2001). Competition Among Foreign and Chinese Agro-Food Enterprises in the Process of Globalization. *International Food and Agribusiness Management Review, 2* (3/4), 437–451.

Wood, A. and Berge, K. (1997). Exporting Manufactures: Human Resources, Natural Resources, and Trade Policy. *The Journal of Development Studies, 34* (1), 35–59.

Wood, A. and Owens, T. (1997). Export-Oriented Industrialization Through Primary Processing? *World Development, 25* (9), 1453–1470.

World Bank (2005). *Food Safety and Agricultural Health Standards: Challenges and Opportunities for Developing Country Exports*. World Bank, Washington, DC.

3 Agro-industry Trends, Patterns and Development Impacts

JOHN WILKINSON[1] AND RUDI ROCHA[2]

[1]Professor and Researcher, CPDA, Universidade Federal Rural do Rio de Janeiro, Brazil; [2]PhD Candidate, Department of Economics, Pontifícia Universidade Católica do Rio de Janeiro, Brazil

Introduction

Agro-industry, understood here broadly as postharvest activities involved in the transformation, preservation and preparation of agricultural production for intermediary or final consumption, typically increases in importance with regard to agriculture and occupies a dominant position in manufacturing as developing countries step up their growth. In all developing countries population growth is becoming predominantly an urban phenomenon, increasing the role of agro-industry in mediating food production and final consumption. While many long-standing commodity exports have declined in importance, 'non-traditional' food exports, especially fruits, horticulture and fish products, and components of the animal protein complex, have become central to developing country exports. Whether looked at from the point of the domestic market or exports, therefore, agro-industry plays a fundamental role in the creation of income and employment opportunities in developing countries.

The agro-processing sector covers a broad area of postharvest activities, comprising artisanal, minimally processed and packaged agricultural raw materials, the industrial and technology-intensive processing of intermediate goods and the fabrication of final products derived from agriculture. The hybrid characteristics and heterogeneous features of the agro-processing sector, ranging from the informal contract relations of poor rural communities to the complex, transnational activities of global players, suggest the need for caution when presenting an empirical overview, which is the main objective of this chapter.[1]

---

[1] The overview initially follows the conventional agro-industry divisions according to the International Standard Industrial Classification (ISIC – aggregating the 3- and 4-digit levels of Revision 3), which includes the main sub-sectors of: (i) food and beverages; (ii) tobacco products; (iii) paper and wood products; (iv) textiles, footwear and apparel; (v) leather products; and

The agrifood sector can be seen as comprising: (i) products for subsistence and local markets (basically root crops); (ii) staples for urban domestic markets (predominantly cereals); (iii) traditional export commodities (coffee, cocoa, tea, nuts, cotton); (iv) components of animal protein diet (dairy products, oils and animal feed) and different meat chains (red meat, pigs, poultry) for both domestic and export markets; (v) fresh or non-traditional products (fruits, horticulture, flowers, seafood/aquiculture); and (vi) differentiated traditional exports (fair trade, organics, origin products), which are now oriented also to domestic markets.

Traditional export commodities are primarily tree crops integrated into multi-crop family farming systems. Non-traditional, fresh products tend to demand greater specialization. The latter are particularly associated with labour-intensive, both in production and postharvest, activities. It is argued that they are particularly beneficial from the point of view of the trade balance, and that they tend to offer greater opportunities for capacity building, given the need to transfer know-how on technical demands related to quality and the nature of market demand (Athukorala and Sen, 1998). Others argue, on the contrary, that cheap land and labour are still the key attractions for investing in developing countries, leading to a 'race to the bottom' as countries and regions bid for investments on the basis of these spurious advantages (Gibbon and Ponte, 2005). Other research suggests that the logistical and quality demands of non-traditional products are leading to a shift away from smallholders to large-scale commercial farms (Dolan and Humphrey, 2000). Traditional export commodities or fresh 'non-traditional' products require special quality attributes, which involve new forms of economic coordination through contracts and supply-chain management. Here the 'global value chain' (GVC) literature is particularly important (Gereffi *et al.*, 2005).

This chapter presents an empirical panorama of the agro-processing sector, selecting the most recent data for each country and identifying dynamic trends whenever possible. The core indicators highlighted will be agro-processing production and value added, contribution to GDP and participation within the total manufacturing sector, the level of formal employment, its gender composition and differences in productivity. With a broader focus on the agrifood system, we also investigate changes in consumption and international trade patterns.

In the concluding section we place our discussion within a broader consideration of global tendencies, briefly focusing in turn on energy, global warming, innovation and the emerging institutional and regulatory context governing global markets. The central implication we draw is that policies for agro-industry should occupy a central position in developing country strategies and that domestic initiatives should now receive special attention. We then indicate the areas that, in our view, should constitute the focus of policy measures.

(vi) rubber products. We will focus particularly on the different divisions of agrifood production/ consumption in order to highlight important features of food-processing, the largest sub-sector, and the centre of the most dynamic changes in the agro-processing sector during the last decades.

Panorama: Core Indicators

Methodology

The methodology adopted in this chapter involves an analysis of a country data set constructed on the basis of the *UNIDO Industrial Statistics Database 2005*, and organized according to the World Bank classification of high-income country (HIC), upper-middle-income country (UMIC), lower-middle-income country (LMIC) and low-income country (LIC). Basic trends in both non-food and food are presented, but the study as a whole focuses on the food sector. Data were also drawn from the *World Development Report* (WDR) *2008* (World Bank, 2007) typology for developing countries, which distinguishes predominantly agricultural countries, developing countries in transformation where the urban economy begins to dictate growth in spite of the numerical superiority of the rural population and urbanized developing countries. Other data sources drawn on include FAO, ILO, the Confederation of European Agro-industries (CIAA), UNCTAD, USDA and World Development Indicators.

For the discussion of the major drivers behind the recent transformation in the agro-industrial sector an extensive literature review was carried out. Of particular relevance here have been the value chain approaches associated with Duke University in the USA, the IDS in England and the DIIS in Denmark; the farm–agribusiness linkages initiatives carried out in all three developing country continents by FAO; the research into non-traditional food exports undertaken by Athukorala and colleagues; the African Regional Working Papers of the World Bank; and research carried out within the framework of CIRAD's cooperation programmes, together with the research coordinated by RIMISP on re-governing markets. We have also benefited from numerous individual studies, in the form of working papers, published articles and books.

Agro-industry production and development impacts

An extended definition of the agro-processing sector, including not only agro-related industries, but also distribution services and trading activities, would roughly account for more than one-third of the GDP of Indonesia, Chile, Brazil and Thailand, and between 20% and 25% of GDP in sub-Saharan countries. The entire food system, including the production of primary goods and commodities, marketing and retailing, would account for more than 50% of the GDP in developing countries (Jaffee *et al.*, 2003, based on World Bank, FAO and UNIDO databases).

In order to gather comparable data within a narrower, more industry-specific perspective, we used only the *UNIDO Industrial Statistics Database 2005*, selecting countries for which data are available on a consistent basis and grouping them according to the World Bank country classification by level of income per capita.

On the basis of this analysis, formal agro-processing participation in the overall gross product corresponds to around 4.3% in LICs (which include

Bangladesh, Ethiopia, Eritrea, India, Mongolia, Senegal and Vietnam) and about 5% in LMICs and UMICs[2] (see Table 1). Considering the importance of artisan production and the informal sector in this activity, particularly in LICs, but generally in the developing world, we can safely interpret this information as heavily underestimating the real picture.

Within manufacturing or production, the agro-processing sector in developing countries occupies a relevant place in overall turnover and value added, particularly for the least- and less-developing countries, though huge heterogeneity may exist among them. Considering the group of LICs analysed here, on average, about 52% of total manufacturing value added corresponds to the agro-processing sector; for the LMICs and UMICs we find figures of, respectively, 36% and 32%. In agriculture-based countries the contribution of agro-processing to total manufacturing is 66%, while in transforming and urbanized countries the figures are, respectively, 38% and 37%.

Based on Jaffee *et al.* (2003) we calculated the ratio of agribusiness share over the agriculture share of GDP for a group of selected countries, which includes a representative sample of sub-Saharan African countries, transforming countries (Indonesia and Thailand), urbanized countries (Latin America and South Africa) and the USA. Agribusiness provides inputs to farmers and connects them to consumers through the handling, processing, transportation, marketing and distribution of agricultural products. According to the WDR (2008), strong synergies can exist between agribusiness, the performance of agriculture and poverty alleviation: efficient agribusiness can spur agricultural growth and a strong link between agribusiness and smallholders can reduce rural poverty. According to FLO (2007), recent trends show that there has been a rapid increase of production value adding via agribusiness opportunities relative to primary agricultural production. Demand from agro-processing increases as does the effective size of the market for agricultural products. Traders and agro-processing firms furnish crucial inputs and services to the farm sector, inducing productivity and product quality improvements, stimulating market growth and innovation throughout the value chains. In this case, the agribusiness/agriculture ratio captures the degree of productive and commercial development of agro-related activities, the sophistication of agro-industrial backward and forward linkages, the capacity level of value adding and market creation, and the importance of distributing and retailing. For agriculture-based countries, for instance, moving the core economic activities from the farm gate to the agro-industrial sector and its services may represent productive diversification and lead to higher levels of productivity and income generation as well as higher shares of non-farm employment in rural areas. Above all, at an aggregate level, this ratio may capture the level of structural transformation currently faced by developing countries, where productivity growth corresponds to a shifting sector composition of economic activity, a fall in the share of agriculture

[2] LIC: Low-income country
LMIC: Lower-middle-income country
UMIC: Upper-middle-income country.

Table 1. Agro-processing participation and the agribusiness/agriculture ratio.

	Year	Agriculture share of GDP[a] (1)	Agribusiness share of GDP[a] (2)	(2)/(1) Ratio	% Agro-processing sector in GDP[b]	% Food-processing and beverages in total manufacturing[b]	% Agro-processing in total manufacturing[b]
Cameroon		0.40	0.17	0.43	–	–	–
Côte d'Ivoire		0.28	0.26	0.93	–	–	–
Ethiopia	2002	0.56	0.30	0.54	0.053	0.49	0.69
Eritrea	2001	–	–	–	0.047	0.45	0.64
Ghana		0.44	0.19	0.43	–	–	–
Kenya		0.26	0.23	0.88	–	–	–
Nigeria		0.42	0.16	0.38	–	–	–
Tanzania		0.32	0.21	0.66	–	–	–
Uganda		0.41	0.23	0.56	–	–	–
Agriculture-based countries		0.39	0.22	0.57	0.050	0.468	0.664
Bangladesh	1998	–	–	–	0.036	0.10	–
Mongolia	2000	–	–	–	–	0.49	–
Vietnam	2000	–	–	–	0.058	0.25	0.41
India	2001	–	–	–	0.022	0.11	0.27
Indonesia	2002	0.20	0.33	1.65	0.066	0.11	0.42
Morocco	2001	–	–	–	–	0.20	–
Philippines	1999	–	–	–	–	0.18	–
Senegal	2002	–	–	–	0.033	0.37	–
Thailand	1998	0.11	0.43	3.91	–	0.25	–

	Year						
Egypt, Arab Rep.	2002	–	–	–	0.022	0.18	0.30
Zimbabwe		0.18	0.21	1.17	–	–	–
Transforming countries		0.16	0.32	1.98	0.040	0.224	0.381
Bolivia	1998	–	–	–	0.056	0.34	0.42
Brazil	2002	0.08	0.30	3.75	–	0.18	–
Bulgaria	2002	–	–	–	–	0.14	–
Argentina	1999	0.11	0.29	2.64	–	0.28	–
Chile		0.09	0.34	3.78	–	–	–
Czech Republic	1999	–	–	–	–	0.09	–
Hungary	2000	–	–	–	–	0.14	–
Mexico	2000	0.09	0.27	3.00	0.037	0.24	0.35
Oman	1997	–	–	–	0.092	0.18	0.23
Russian Federation	2002	–	–	–	–	0.19	–
South Africa	1996	0.04	0.16	4.00	–	0.15	–
Uruguay	2000	–	–	–	0.054	0.31	0.49
Urbanized countries		0.08	0.27	3.32	0.060	0.203	0.375
USA		0.01	0.13	13.00	–	–	–
Unweighted average							
LICs					0.043	0.32	0.52
LMICs					0.055	0.20	0.36
UMICs					0.051	0.20	0.33

aSource: Jaffee *et al.* (2003) for agriculture and agribusiness share of GDP. Agribusiness combines the value added for agro-related industries and that of agricultural trade and distribution.
bUNIDO *Industrial Statistics Database* (*2005*) for agro-processing data with respective year.
Unweighted averages consider all information available in each column.

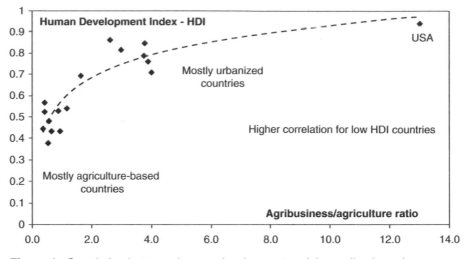

Figure 1. Correlation between human development and the agribusiness/agriculture ratio.

and increasing transfers of capital and labour from agriculture towards expanding agro-industrial and related service sectors.

In the USA agribusiness contributes 13 times more to GDP than pure agricultural activities. In urbanized developing countries, following the WDR typology, this ratio remains at 3.3, whereas in transforming countries it falls below 2 and in agriculture-based countries it is only 0.6. More fundamentally, and not surprisingly, this ratio is highly correlated with basic measures of socioeconomic development. Low indices of human development are directly related to low ratios of agribusiness-to-agriculture development. Socio-economic catch-up, on the other hand, can be highly and positively correlated with levels of economic growth passed on from agriculture to agro-related manufacturing and service activities (Figure 1).

According to the WDR, growth in rural non-farm employment is in many cases an important factor in rural poverty alleviation and remains closely linked to improvements in agriculture. Rural trade and transport, often of food, would represent about 30% of rural non-farm employment.[3] The direction of causality, however, is conditional on specific circumstances. Some estimates for rural China highlight the effects of growth on farming rather than on non-farming activities, with less evidence of reverse linkages. On the other hand, with urbanization becoming an almost generalized worldwide trend, growth in rural non-farm employment occurs independently of agriculture performance. When capital and products are mobile, investors search for low-wage opportunities in areas that have not increased their incomes through higher agricultural productivity. There are also generally areas that are closer to primary agricultural inputs.

---

[3] The report identifies a high correlation between the ratio of food-processing to agricultural value added and income per capita for a sample of developing countries.

For instance, urban overcrowding and higher labour costs have stimulated urban-to-rural subcontracting in East Asia, both for domestic consumption and for export. In this case, although relatively few of the poor gain access to non-farm jobs in rural areas, higher labour demand would indirectly put upward pressure on agricultural wages.

Production: basic stylized facts within agro-processing

Within agro-industry, food-processing and beverages are by far the most important sub-sector in terms of value added, accounting for more than 50% of the total formal agro-processing sector in LICs and LMICs, and more than 60% in UMICs (see Figure 2). For the African countries included in Table 1, Ethiopia, Eritrea and Senegal, food and beverages represent more than 70% of agro-industry value added and roughly 30–50% of total manufacturing. On the one hand, tobacco and textiles have played an important role in Asian and Middle-Eastern countries, while wood, paper and rubber production are heavily concentrated in Asian countries. Leather products, on the other hand, represent only a marginal share in the total agro-processing value added. According to FAO (2007), throughout the past 25 years, the shares of global manufacturing value addition for food, beverages, tobacco and textiles (which are the main agro-industry manufacturing product categories tracked by UNIDO) generated by developing countries have almost doubled. For textiles, developing countries accounted for 22% of manufacturing value added in 1980, but more than 40% in 2005. The increase was the greatest for tobacco, reaching 44% of global value addition in 2005. In order to focus on sub-sectors that make up almost

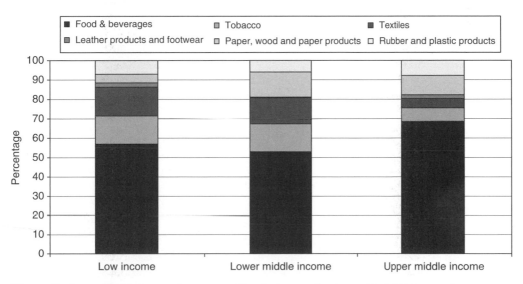

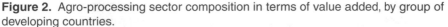

Figure 2. Agro-processing sector composition in terms of value added, by group of developing countries.

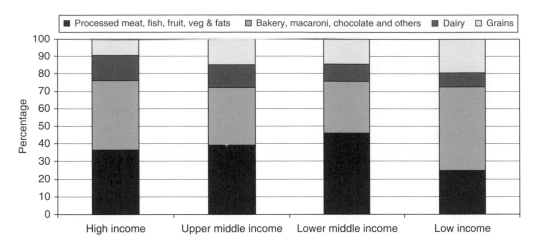

Figure 3. Food-processing sub-sector composition in terms of value added.

all the agro-processing value added in LICs, the chapter will concentrate on textiles, tobacco and, above all, food and beverages.

Considering specifically the food-processing sub-sector, Figure 3 shows that processed meat, fish, fruits, vegetables and fats, and bakery, macaroni, chocolate and others represent together 70–75% of the total value added. Grains are relatively more important for LICs and dairy products for HICs. Food industries in emerging countries are undergoing considerable expansion, particularly in Latin America and Asia. Brazilian and Chinese food production recorded double-digit growth rates (16% and 22%, respectively) from 2001 to 2004 (CIAA, 2006). According to FAO (2007), there are large differences among developing regions in the distribution of formal sector agro-industry value added. Latin American countries accounted for nearly 43% of food and beverages value addition in 2003 and countries of South and South-east Asia for 39%. African countries, however, contributed less than 10%.

The textile, clothing, leather and footwear sub-sectors are among the most globalized activities of the agro-processing sector. As a result of this globalization process, the speed of which has increased in the last 2 decades, world distribution of production, trade and employment has changed dramatically over the years, and is likely to change even more given the recent phasing out of the Agreement on Textiles and Clothing (WTO, 2007).

Labour productivity

Productivity levels within agro-industry are heterogeneous, ranging from low for textiles and wood products to extremely high for tobacco products. When compared to agricultural or general manufacturing standards, agro-industry approaches industrial averages or even reaches relatively higher productivity levels, as is the case for food and beverages in LICs and MICs (Table 2).

Table 2. Labour productivity in agro-industry (value added over number of workers, average in current US$).

	Year[a]	Food & beverages[a]	Tobacco products[a]	Textiles[a]	Wood products[a]	Paper and paper products[a]	Rubber and plastic products[a]	Total manufacturing[a]	Agricultural productivity[b]
Bangladesh	1998	2,939	40,192	862	1,437	2,007	1,967	2,066	308
Ethiopia	2002	7,756	23,674	1,215	–	3,178	6,162	4,696	149
India	2001	3,192	1,754	2,667	1,801	4,631	5,525	5,053	381
Mongolia	2000	2,695	–	2,580	1,059	759	–	1,619	684
Senegal	2002	11,952	35,459	4,863	10,569	13,113	9,805	13,843	249
Vietnam	2000	3,146	–	–	–	–	–	2,841	290
Unweighted average – LICs		5,280	25,270	2,437	3,716	4,737	5,865	5,020	344
Brazil	2002	18,047	59,865	10,368	7,730	34,801	13,587	19,559	2,790
Bulgaria	2002	2,782	4,969	2,607	1,838	2,399	3,093	2,864	6,313
Indonesia	2002	6,206	9,553	4,532	5,117	16,355	3,419	7,056	556
Morocco	2001	11,523	255,815	7,001	6,061	17,561	9,910	11,657	1,515
Philippines	1999	26,423	56,854	7,310	5,572	15,053	10,942	16,065	1,017
Thailand	1998	10,499	1,996	5,484	3,261	9,929	6,043	8,276	586
Egypt	2002	6,028	9,722	1,878	2,812	9,047	3,891	6,883	1,975
Unweighted average – LMICs		11,644	56,968	5,597	4,627	15,021	7,270	10,337	2,107
Argentina	1999	33,255	71,516	21,468	15,909	35,255	29,750	33,843	9,272
Czech Republic	1999	10,872	–	6,864	7,539	11,640	11,595	9,758	4,564
Hungary	2000	9,942	38,988	5,412	5,374	14,208	10,810	11,436	5,080
Mexico	2000	39,964	254,613	14,779	12,259	43,248	23,767	41,156	2,704
South Africa	2001	10,789	–	–	–	–	–	9,329	2,391
Uruguay	2000	25,698	311,406	20,412	18,099	34,452	19,356	35,651	6,743
Unweighted average – UMICs		21,753	169,130	13,787	11,836	27,761	19,056	23,529	5,126

[a] UNIDO Industrial Statistics Database 2005 for agro-processing data with respective year, in current US$.
[b] Source: WDR (2008), data for 2001–2003, US$ 2000 prices.

Table 3. Productivity in the food-processing sector (value added per worker, average in current US$). (Based on the *UNIDO Industrial Statistics Database 2005*. Countries within each category are the same as presented in Table 1.)

	Processed meat, fish, fruits, vegetables and fats	Dairy	Grains	Others: bakery, macaroni, chocolate, etc.	Food-processing sector	Total manufacturing
LIC	3,830	9,418	6,388	4,395	4,937	4,804
LMI	15,941	21,090	15,587	10,605	15,083	15,694
UMI	18,023	21,855	29,308	17,919	18,296	23,076
HIC	46,675	71,439	87,569	61,433	55,408	57,738

The fact that productivity levels for food processing are above the manufacturing average not only complements the patterns identifying the sector as one of the largest industrial activities in LICs and MICs in terms of value adding, but also confirms it as one of the most efficient economic sectors in least-developed countries and the one that pushes the manufacturing sector towards higher levels of technical capabilities and value-adding achievements. Table 3 shows that dairy products present the highest labour productivity levels in LICs and LMICs, while grains occupy this position in UMICs, as well as in HICs.

Employment, informality and gender composition

According to the ILO, on average, 60% of the workers in the food and beverage industry in developing countries are employed in the informal economy, occupying jobs that are often precarious in terms of social protection. A major problem here is the absence of comparable statistics. Following the ILO estimates and considering only countries where official statistics are available and on the basis that data cover only the formal economy, we may have some 22 million people employed globally in the food and drink industries.[4] Both the ILO and the UNIDO statistics confirm a decline of employment in the food and drink industry in many developed countries, often as a result of the relocation of processing operations to developing and transition economies. On the other hand, there has been strong growth in employment in some developing countries (Thailand, Mexico, the Philippines) while other countries (e.g. South Africa) have experienced a sharp fall in recent years (ILO, 2007).

Many high-value agrifood and non-food chains are characterized by increasing levels of female participation (Dolan and Sorby, 2003). In the Dominican Republic, women comprise roughly 50% of the labour force employed in horticulture processing (Raynolds, 1998). In Mexico, 89–90% of employees in packaging are women (Barrón, 1999). In Kenya and Zambia, over 65% of workers in horticulture pack-houses and farms are women (Barrientos *et al.*, 2001;

---

[4] Based on LABORSTA. More information at: http://www.ilo.org/public/english/dialogue/sector/sectors/food/emp.htm.

Dolan and Sutherland, 2002). Women represent 91% of horticultural workers in Zimbabwe (AEAA, 2002). In Chilean fruit production, female employment increased almost 300% between 1982 and 1992, an impressive pattern when compared to a national growth rate of 70% for the female labour force (Barrientos, 1997). In Brazil, 65% of workers in fruit production were women in the mid-1990s (Collins, 2000). In horticulture, the highest levels of female participation have been documented in the cut flower sector. In Tanzania, Ecuador, Kenya and Uganda, women represent respectively 57%, 70%, 75% and 85% of workers (Blowfield *et al.*, 1998; Palan and Palan, 1999; Asea and Kaija, 2000; Dijkstra, 2001). Poultry processing is another labour-intensive activity and absorbs high levels of female workers. In 2000, 80% of employees in Thailand's Cargill subsidiary, Sun Valley, were women – a proportion similar to that found in other poultry processors (Lawler and Atmananda, 1999).

Dolan and Sorby (2003) argue that flexible labour follows gender-based patterns, with women allocated to more vulnerable forms of work (casual, temporary and seasonal), and men concentrated in the fewer permanent jobs. In most activities, there is strong gender segmentation in both production and processing, reinforced by prevailing gender stereotypes. Women are considered to have higher skills for tasks requiring manual dexterity and patience, such as harvesting, sorting, grading, de-boning and packaging. Men, for their part, are seen to have superior physical strength, supervisory capacity and mechanical skills.

According to ILO (2005), the participation of female workers in the clothing and textile industries is above the manufacturing average and significantly higher for the clothing sector. Women are often young and low skilled. The share of female employment in clothing is considered to be more than 89% in Cambodia, 80% in Bangladesh and 82% in Sri Lanka. Female participation in India and Turkey is below 50%, while in Guatemala it reaches around half of total employment. Female participation in textiles is generally lower, below 50%, with the exception of Cambodia (76%) and Sri Lanka (61%).

The role of the rural non-farm economy[5]

According to the WDR around 75% of the poor in developing countries live in rural areas, 2.1 billion living on less than US$2 a day and 880 million on less than US$1 a day. Although agriculture remains at the productive centre of most rural economies, labelling them as purely agricultural is clearly inaccurate. For developing countries as a whole, non-farm earnings account for 30–45% of rural household income. This is not only a large share in absolute terms but increases over time, constituting a complement to agricultural wages and an instrument for risk diversification and the evening out of consumption patterns. With low capital requirements and undemanding local marketing channels, the rural non-farm economy offers opportunities for poor households, small-scale

---

[5] Based mainly on Haggblade *et al.* (2005).

farmers and other smallholders, representing a potentially important instrument for poverty alleviation in rural areas.

Haggblade *et al.* (2005) argue that the rural non-farm economy plays an important role in the process of structural transformation, during which the share of agriculture in national output declines and transfers of capital and labour drive a corresponding rise in manufacturing and services, particularly those related to agro-industry. Here, therefore, we have a key for the understanding of many of the processes driving overall economic growth and poverty reduction in least developed countries.

The rural non-farm economy is a heterogeneous set of trading, agro-processing, manufacturing, commercial and service activities, ranging from part-time artisanal entrepreneurs to large-scale industrial plants operated by multinational firms. Some clear compositional patterns can be identified spatially, with home-based cottage industries, small-scale retailers and basic farm equipment repair services predominating in rural areas while factories, traders and transport facilities, public administration offices, schooling, health clinics and other services are concentrated in small towns.

In terms of sector composition, rural industries account for only 20–25% of rural non-farm employment (see Table 4), consisting mostly of occupations in agro-industries. Indirectly, however, other activities such as commerce and retailing, construction, transport and trade are typically associated with agro-related manufactures and agribusiness.

Agriculture evidently has a direct influence on the size and structure of the rural non-farm economy since it accounts for the largest share of employment, value addition and raw material supplies. Haggblade *et al.* (2005) discuss two main scenarios for the dynamic relation between agriculture and the rural non-farm economy. On the one hand, where new technologies and modern farm inputs are available there would be productivity surpluses and increasing opportunities for trade, capital accumulation and value addition. A dynamic agriculture fuels economic growth in the rural non-farm sector through a number of linkages. It requires inputs, such as seeds, fertilizer, credit, machinery, marketing and processing facilities, which create an increasing demand for non-farm enterprises that provide such goods and services. Still, some non-farm activities, initially undertaken by farm households for self-consumption, spin off as full-time commercial enterprises, while others, particularly labour-intensive household manufacturing, die out in rural areas, displaced by imports from cheaper urban factories. This demise of low-productivity household manufacturing as well as changes in consumption patterns towards modern urban-related habits would, in part, explain why employment in services and commerce often increases faster in rural areas than local manufacturing and small-scale agro-processing plants.

Consumption patterns of processed food and beverages

As presented previously, processed food and beverage products account, on average, for almost 50% of total agro-processing value added in developing countries, and occupy an even more relevant place in the overall manufacturing

Table 4. Composition of rural non-farm employment by region. (From Haggblade *et al.*, 2005.)

| | Non-farm share of rural workforce | Women's share of rural non-farm employment | Rural non-farm employment shares | | | | |
			Manufacturing	Trade & transport[a]	Financial and personal services[b]	Construction, utilities, mining and others[c]	Total rural non-farm
Africa	10.9	25.3	23.1	21.9	24.5	30.4	100
Asia	24.8	20.1	27.7	26.3	31.5	14.4	100
Latin America	35.9	27.5	19.5	19.6	27.3	33.5	100
West Asia and North Africa	22.4	11.3	22.9	21.7	32	23.2	100

[a]Trade and transport include wholesale and retail trade, transport and storage.
[b]Other services include insurance and community and social services.
[c]Others include quarrying and other non-classified activity.

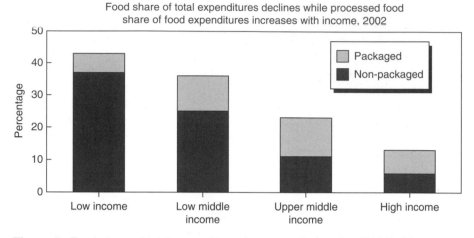

Figure 4. Food share of total expenditures by group of countries (2002). (From Gehlhar and Regmi, 2005, based on Euromonitor.)

value added of LICs. We now focus on consumption trends, the primary driving force of the agro-processing sector and its most dynamic feature. The following discussion presents recent trends in consumption patterns with particular attention given to the changing composition and growth rates of consumption in developing countries.

Global sales of food and beverages were estimated at US$4 trillion[6] in 2002, around 80% of which corresponded to processed food and beverages (US$3.2 trillion) with over 40% accounted for by the food service sector. Within the retail sector US$531 billion of food sales corresponded to fresh products, while US$1.7 trillion were made up of processed foodstuffs – about US$1.1 trillion of which was packaged food with beverages amounting to US$641 billion (Gehlhar and Regmi, 2005). Expenditure on processed food and beverages approached US$4.8 trillion in 2007, which is 57% higher when compared to that of 2001, indicating a recent average annual growth rate of 7% (ILO, 2007).

Following the calculations of Gehlhar and Regmi (2005) based on Euromonitor (2003), food consumption in HICs accounted for over 60% of packaged food sales in the world and half of total food expenditures. In 2002, per capita retail sales of packaged food in these countries were US$979, more than 15 times the value found in the case of the LICs, which was some US$63. In spite of this very low level of expenditure we find that the share of food in total expenditures is more than 40% in LICs, mostly on non-packaged food (see Figure 4). With rising incomes, relative expenditure on food share declines, although the share of processed food in total food expenditures increases.

According to FAO (2001) (*Food balance sheets 1999–2001*), large differences in calorie consumption patterns between developed and developing

---

[6] One trillion = 1,000,000,000,000.

countries are also found – 3261 versus 2675 calories per capita per day, respectively. As regards diet composition, meat, milk and dairy products account for only 5% of total calories consumed per day in developing countries, while they amount to around 19% in the developed world. Cereals, on the other hand, account for 53% of calories consumed in developing countries, against 31% in developed countries (see Figure 5). Also, according to Gehlhar and Regmi (2005, p. 10), the number of products purchased at retail outlets is greater for wealthier countries, reflecting the increased demand for variety as incomes increase. The top five product categories account for 71% of processed food retail sales for Mexico and 74% for India, but only 48% for the USA and 47% for the UK. The authors point out that, as the demand for processed foods and beverages (soft drinks) is also driven by the demand for higher quality, the items consumed by countries at different income levels reflect different levels of demand for services embodied in the products. For example, ready-to-eat meals account for about 4% of total retail sales in the USA and the UK, but only 0.06% in Mexico and 0.55% in China. On the other hand, intermediate products, such as fats and oils, while accounting for over 7% of total processed food retail sales in India, 13% in Indonesia and 5% or more in many developing countries, account for less than 2% of retail sales in HICs.

In spite of relatively low levels of consumer expenditure on processed food and beverages in developing countries, these figures are changing rapidly. Between 1996 and 2002, while retail sales of packaged foods have grown at about 2% or 3% per year in HICs, they have grown much faster among developing countries, ranging from 7% in UMICs to 28% in LMICs and 13% in LICs. In developed countries, growth in food consumption and beverages is expected to rise mainly from slow rates of population growth rather than from increases in per capita consumption. On the other hand, as a result of increases in population, income and per capita food consumption, developing countries are expected to account for most future increases in processed food consumption. In this context, the high processed food consumption growth rate observed in MICs, mainly driven by urbanized economies, may soon be paralleled by China, Thailand, the Philippines, Indonesia, Vietnam and India.

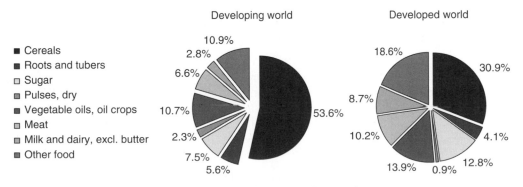

Figure 5. Consumption: composition of calories per capita per day.

Income and population growth are immediate determinants of increasing processed food consumption and the upgrading of diets towards variety and quality. Population growth in a context of urbanization brings changes in food consumption based on distance and time constraints, placing a premium on food preservation and convenience. In developing countries, consumers whose diets have been low-value, carbohydrate-rich cereals increase their expenditures on higher-value meats, fruits and vegetables. Consumers in developed countries and the middle classes in developing countries are shifting their diets towards foods that reflect not only increases in the nutrient value of the food basket, but also the value-added services embodied in the products (Gehlhar and Regmi, 2005).

Increasing sales of ready-to-eat meals, convenience food and food services have also been promoted by further demographic and social changes, including the increasing participation of women in the labour market, the ageing of the population and the rising importance of single-person households. International tourism and more culturally diverse societies, as a result of international migration, are also leading to changes in food tastes and higher demand for ethnic products (ILO, 2007). Processed foodstuffs and soft drinks carry a component of prestige, which makes them attractive to consumers. Preoccupations with health issues and food safety further the demand for modified (diet, enriched) products. Also, Gehlhar and Regmi (2005) highlight the importance of increases in income on the acquisitions of refrigerators (which may lead to greater household purchases of perishable food products and frozen, ready-to-eat foodstuffs) and microwave ovens (increasing the consumption of prepared foods and retail sales of ready meals). While concerns for health have been a major innovation driver (diet, light) in the food industry, the correlation between increased consumption of processed foods and obesity and food-related illnesses in both developed and developing countries has drawn attention to high levels of fats, sugar and oils in processed and, especially, convenience foods. In addition, recent research suggests that these levels increase in the case of cheaper brands, reinforcing the association between poverty and obesity.[7]

International trade: agricultural products, processed food and other high-value agrifood items

In terms of international trade many important trends have occurred during the last decades in the agricultural sector. We will mainly focus on points of relative consensus in relation to recent trends. First, despite natural comparative advantages and due primarily to protective trade regimes and distorted tariffs in developed markets, developing countries have still the same market share in world agricultural trade as they had in the 1980s. Second, trade composition has changed dramatically, as a result of stagnant markets for traditional commodities and expanding demand for fruits, vegetables, fisheries and beverages.

---

[7] These issues were discussed at length at the FAO Technical Workshop: *Globalization of food systems: impacts on food security and nutrition*, Rome, October 2003.

Other relevant compositional trends include the increasing share of processed products in agricultural trade and growing South–South trade flows. Third, we highlight the small share of traded processed food compared to overall world sales. An important corollary here is the strong domestic bias of food consumption, which favours FDI rather than trade.

According to Aksoy's (2005) calculations, the market share of developing countries in world agricultural trade in 2000–2001 was around 36%, a slightly lower figure when compared to that of 1980–1981 (see Table 5). This performance would be worse without the increase in trade among developing countries during the 1990s. Agricultural trade among these countries has been characterized by its heterogeneity. Trade expanded in the case of LICs, primarily driven by imports from other developing countries. As a group, MICs have performed worse, although selected countries are becoming major exporters, such as Argentina, Brazil and Thailand. The increasing presence of Argentina and Brazil in export markets is particularly notable. Brazil's export performance is mostly concentrated on sugar, oilseeds and meats, while Argentina's also covers cereals and dairy products. Other emerging exporters in developing and transition economies include Russia and the Ukraine for coarse grains, Vietnam and Thailand for rice, Indonesia and Thailand for vegetable oils and Thailand, Malaysia, India and China for poultry (OCDE and FAO, 2007). On the other hand, UMICs in East Asia are becoming major importers. Nevertheless, flows among developed countries still represent the dominant tendency in global agricultural trade, accounting for 50% of the total, with 60% of this being conducted within trading blocs such as the European Union (EU) and NAFTA.

With the recent explosive growth of China, India and other large developing countries, there has been a sharp increase in commodity exports from middle-income developing countries based on the animal protein complex (meats, animal feed and, more recently, dairy products), leading to a greater proportion of South–South flows in world trade. At the same time, South–South trade has involved an increase in imports (especially poultry) into the LICs, which threatens the ability of these countries to develop their domestic agro-industrial base in this sub-sector. In fact, trade statistics reveal that since the mid-1980s, in South–South trade, primary commodities have played a

Table 5. Shares of developing and developed countries in world agricultural exports. (From Aksoy, 2005, based on COMTRADE.)

	Developing countries			Industrial countries		
	1980–1981	1990–1991	2000–2001	1980–1981	1990–1991	2000–2001
To developing countries	13.4%	10.5%	13.7%	18.9%	14.5%	15.6%
To industrial countries	24.3%	22.4%	22.4%	43.4%	52.5%	48.3%
Total	37.8%	33.0%	36.1%	62.0%	67.0%	63.9%

more important role than in South–North trade. In 2003, the major exporters of agricultural products within South–South trade were China (11.5% of total agricultural South–South exports), Argentina (10.6%), Brazil (10.2%), Malaysia (9.6%), Thailand (8.2%), Indonesia (6.5%) and India (5.5%). The top ten exporters accounted for more than 70% of total agricultural South–South exports. The major importers were China (18%), Hong Kong (7.4%), Republic of Korea (7.2%), India (6.1%), Malaysia (4.2%) and Brazil (3.9%), with the top ten importers accounting for 60% of total agricultural imports (UNCTAD, 2005).

Although aggregate export shares have been steady for developed and developing countries throughout the last 2 decades, agricultural trade has been marked by dramatic commodity composition changes. Exports of high-value agricultural and processed food products, such as fresh and processed fruits, vegetables and fisheries, have expanded significantly, stimulated by changing consumer tastes and advances in production, transport and other supply chain technologies. Trade performance in traditional commodities, however, has declined as has the share of these products in developing country agricultural exports (Henson, 2006).

The change in the composition of agricultural exports in developed and developing countries in the last 2 decades is presented in Table 6. Non-traditional and other processed products expanded their participation in global exports, from around 31% in 1980–1981 to 50% in 2000–2001, while the share of tropical and temperate products declined from 69% to 50%. Among the worst performances, textile fibres (mostly cotton) and sugar and confectionery have lost 50% of their shares and the global participation of grains has declined 40%. This result is fuelled by even larger losses in developing country exports. Fisheries and beverages, on the other hand, have presented the highest growth rates. According to Aksoy (2005), traditional products such as coffee, cocoa and tea, which have received most of the attention in the literature, now account for less than 20% of the exports of developing countries. Given current agricultural export patterns, more attention should now be paid to the expanding trade among developing countries, particularly in temperate zone products such as milk, grains and meats.

According to FAO (2004a, p. 14), imports by developing countries increased rapidly during the 1970s, grew more slowly during the 1980s and accelerated again throughout the 1990s. The food trade surplus of US$1 billion for developing countries became a deficit of more than US$11 billion during the period. In particular, UNCTAD (2006) points out that, although food exports constituted 13.6% of the least developed countries' total exports in 2000–2003, the overwhelming majority of these countries were net food-importing countries, with food imports averaging almost one-fifth of their total imports. Much of this deficit corresponds to processed food imports, given that in developing countries around 65–70% of total food imports in 2002 were for processed products. According to FAO (2004a), the economic performance of individual developing countries played an important role in determining how fast they increased their food imports during the 1990s. Countries that recorded strong economic growth increased food imports more quickly. Rapid growth in the agriculture sector had the opposite effect. Where agricultural value added per capita grew more quickly, food imports generally did not. These trends may

Table 6. The structure of agricultural exports in world trade (percentage of total world trade). (From Aksoy, 2005, based on COMTRADE.)

	Developing countries exports			Industrial countries exports			World exports		
	1980–1981	1990–1991	2000–2001	1980–1981	1990–1991	2000–2001	1980–1981	1990–1991	2000–2001
Tropical products									
Coffee, cocoa and tea, raw and processed	18.3	11	8.5	2.5	2.9	3.6	8.5	5.6	5.4
Nuts and spices	2.4	2.7	2.8	0.7	0.7	0.8	1.3	1.3	1.5
Textiles, fibres	8	6.2	3.3	4.5	3.9	2.6	5.9	4.7	2.8
Sugar and confectionary	10.5	4.6	4.3	3.9	2.8	2.3	6.4	3.4	3.1
Subtotal	39.2	24.4	18.9	11.6	10.3	9.3	22	14.9	12.7
Temperate products									
Meat, fresh and processed	7.2	8.3	6	14.8	15.7	15.4	11.9	13.2	12
Milk and milk products	0.3	0.7	1.1	7.9	7.9	7.6	5	5.5	5.2
Grains, raw and processed	9.3	4.9	7	21.6	13.8	11.6	16.9	10.9	9.9
Animal feed	7.5	7.9	8.5	7.7	5.1	5.3	7.7	6	6.4
Edible oil and oilseeds	4.6	5.7	5.5	4.8	4.4	4.4	4.7	4.8	4.8
Subtotal	28.8	27.5	28.1	56.9	46.8	44.2	46.3	40.4	38.3
Seafood, fruits and vegetables									
Seafood, fresh and processed	6.9	15.9	19.4	5.5	8.2	8	6	10.8	12.2
Fruits, vegetables and cut flowers	14.7	22.2	21.5	13.1	17.2	17.3	13.7	18.9	18.9
Subtotal	21.6	38.2	41	18.7	25.5	25.4	19.8	29.7	31
Other processed products									
Tobacco and cigarettes	2.6	3.1	3.3	3	4.2	4.8	2.8	3.8	4.2
Beverages, alcoholic and non-alcoholic	1.1	1.8	3.6	6.9	9.5	11.5	4.7	6.9	8.6
Other products and processed food	6.7	5	5.2	3	3.8	5	4.4	4.2	5.1
Subtotal	10.4	9.9	12.1	12.8	17.5	21.2	11.9	15	17.9
Total	100	100	100	100	100	100	100	100	100

Table 7. Share of food and processed food in trade (in percentages: unweighted average). (From Wilkinson and Rocha, 2006, based on FAO *Statistical yearbook 2004* (FAO, 2004a)).

	LIC		LMI		UMI		HIC	
	1989/1991	2002	1989/1991	2002	1989/1991	2002	1989/1991	2002
Share of food in total imports	17.4	16.3	13.3	11.7	10.4	9.2	7.4	6
Share of food in total exports	19.2	14.2	22	13.7	22.8	14	6.2	4.6
Share of processed food in total imports	12.8	11.3	8.3	7.4	6.8	6.4	4.4	3.9
Share of processed food in total exports	5.2	4.5	14.3	8.9	10.7	8.2	3.7	3.1
Share of processed food in food imports	72.3	67.7	60.3	63.6	63.3	67.9	61.1	67.3
Share of processed food in food exports	35.6	38.9	54.5	61.6	53.3	63.6	62.3	71.3

Figures for 1989–1991 do not include the former USSR.

suggest that, although an export-led strategy may be appropriate for selected individual developing countries, domestic variables play an important role in preventing food trade deficits.

The increasing importance of processed agricultural products is a long-term trend observed across regions and countries. The share of final agricultural products in global agricultural trade increased from 27% in 1980–1981 to 38% in 2000–2001, varying from 10% in LICs to 45% in HICs. According to Wilkinson and Rocha (2006), the share of processed food in food exports rose between 1989–1991 and 2002 in all country groups (see Table 7). For HICs, this share increased from 62% to 71%, while figures for UMICs (53–64%) and LMICs (54–62%) have followed an even more pronounced upward trend.[8] A closer look at the data reveals, however, that, in reality, there are only a few UMICs and LMICs responsible for a large share of global food-processing exports, i.e. Argentina, Brazil, Chile, Indonesia, Malaysia, Thailand and Turkey.

Despite the increasing share of processed food in global agricultural trade, only 10% of processed food sales globally are traded products. Although consumer demand for processed food continues to grow globally, growth in

[8] Data from the FAO *Statistical yearbook 2004*.

processed food trade has generally stalled since the mid-1990s, while the share of food in total trade has fallen. According to Athukorala and Jayasuriya (2003), the share of processed food in total exports has globally dropped from 8.5% in 1970 to 6.5% in 1990 and 5.8% in 1999 – when processed food exports amounted to US$212.6 billion from developed countries and only US$81.8 billion from developing countries.

In 2006, exports of textiles and clothing were worth US$530 billion, about 4.6% of total world exports (WTO statistics, 2008). Around 59% corresponded to clothing exports, which make up an increasing share of total textile and clothing trade. In terms of development impacts, there are two important features. First, developing countries account for roughly 50% of world textile exports and almost 75% of world clothing exports. Second, as shown in Table 8, in a number of developing countries, textiles and clothing account for the major share of exports, reaching 75–85% of total trade in the case of Pakistan and Bangladesh (ILO, 2005).

With regard to imports, for decades textiles and clothing were subject to the extensive use of quotas by the major importing countries. As stated in the Agreement on Textiles and Clothing (ATC), this system was gradually phased out to 2005. According to the WTO Report (2007), structural changes in the world trade of textiles and clothing continue apace. While China's exports continued to gain market share, exporters from developed countries and those from advanced developing economies in East Asia have lost market share, together with major suppliers from Central America and the Mediterranean region. Some smaller suppliers have expanded their textiles and clothing exports even faster than China and the share of least developed countries in imports by the USA and the EU increased sharply in 2006.

Table 8. Share of textiles and clothing in total exports for selected countries (in percentages, 2003). (From ILO, 2007, based on United Nations Commodity Trade Database – COMTRADE.)

	Textiles	Clothing	Total
Macao, China	0.90	89.90	90.80
Bangladesh	8.70	76.50	85.20
Pakistan	47.70	26.30	74.00
Hong Kong, China	4.90	52.50	57.40
Mauritius	4.20	52.60	56.90
Sri Lanka	4.00	51.60	55.60
Nepal	16.50	34.50	51.00
Morocco	1.50	32.50	34.00
Macedonia, FYR	3.20	30.00	33.20
Madagascar	2.30	30.80	33.10
Turkey	11.00	21.70	32.70
Romania	2.60	23.20	25.70
China	6.30	11.90	18.20
Cambodia	–	80.00	–
Guatemala	–	42.00	–

Complementary Stylized Facts

Foreign direct investment

Foreign direct investment (FDI) has grown much faster than trade in the last 2 decades (Senauer and Venturini, 2004). According to UNCTAD (2006), in 2004 world FDI inward flows regarding the food, beverages and tobacco industry were estimated at US$278 billion, an impressive figure when compared to total world food exports for that year, estimated at US$630 billion (*WTO Statistics Database* (WTO, 2008)), or to total agricultural products exports of US$786 billion.

Only a minor share of world agrifood FDI inward flows, around 14% in 2004, has been directed to developing countries but their impact on host food systems has been profound. Food and beverage manufacturers, long present in many developing countries, are rapidly expanding their operations, and exports now represent only a partial strategy for gaining market share. The choice between exports and FDI has been subject to much discussion and will be taken up in more detail in the next section. A distinction should be made between primary processing and final food activities. The former tend to be located close to the raw material, whether this is in the rural areas, in the case of commodity exports, or by the port, in the case of imports. Final foods, on the other hand, tend to be manufactured close to the consumer market, given the specificity of food regulations and eating habits. FDI has been particularly attracted by the growing urban food markets of developing countries.

Non-traditional products for export, based on the advantages of location combined with the demand for freshness, have created a new dynamic to the extent that postharvest activities have to be carried out *in situ*. While this has led to flows of FDI, it has also given rise to a hybrid form, which is neither trade nor FDI and has been described as the consolidation of Global Value Chains (GVCs), often governed by large-scale retail. Within this dynamic the action of many autonomous agents is coordinated by the demands of the end buyer, who may or may not engage in direct investments but will specify and closely monitor the conditions of production. The literature on GVCs will be discussed in more detail in the following section.

The participation of small and medium enterprises (SMEs)
and industrial concentration trends

A dynamic agribusiness sector linking farmers to consumers can be a major driver of growth in the agricultural and the rural non-farm sectors, particularly offering opportunities for the rural poor. Market structure trends and the role assigned to small-scale operators, however, will be crucial throughout this process.

Agribusiness enterprises are mostly small scale, located in rural towns and headed by households with other complementary sources of income. Medium and large firms are mainly urban based, taking advantage of economies of scale and better infrastructure, while large enterprises are often vertically and horizontally integrated multinational corporations. The term SME encompasses a collection of firms with wide differences in organizational and marketing capabilities and technology, but which are mostly labour-intensive. In developing countries, although SMEs have typically operated on an informal basis, incurring high transaction costs and suffering from a lack of scale, they account for a large share of the number of firms and jobs, contributing a significant share of total value added to the agro-industrial sector.

Experiences in Brazil, Chile, Kenya, Mexico, South Africa, Taiwan and Thailand have demonstrated the potential of agro-based SMEs for employment generation, value adding, food security, poverty alleviation, improvement of farm and rural non-farm income and the living standards among the rural poor more generally. In Africa, where a weakening of public services has resulted in some dysfunctional input and output markets and a partial breakdown in the delivery of agricultural services to small-scale farmers, local agro-enterprises are increasingly filling crucial institutional gaps, particularly for commercial crops (Freeman and Estrada-Valle, 2003).

On the other hand, increasing agribusiness concentration and market displacement of smallholders may hamper the poverty impact of agro-industries development. The structure of agribusiness has changed significantly and its performance has been highly dynamic, particularly enhanced by changing demand patterns and rapid technological, organizational and institutional innovations. According to the World Bank (2007), the sector faces two major challenges when considering development impacts – market forces do not of themselves ensure competitiveness; nor do they guarantee smallholder participation – both of which are essential if agricultural growth is to be linked to overall development and rural poverty reduction.

Scale economies and the globalization of agro-processing markets have consolidated multinational operations along the different GVCs. Food-processing firms are integrated backward into primary product handling and forward into supply chains dominated by global retailers, such as increasingly bypassing smallholders in traditional and local markets. In addition to concentration within sectors, strategic alliances across sectors are becoming common, as between the genetics inputs industry and primary agricultural processing.

With higher industry concentration market competitiveness may decline and can lead to higher spreads between what consumers pay and what producers receive. Three companies control more than 80% of the world market for tea. Some 25 million farmers and farm workers produce coffee, but the international coffee traders have a CR4[9] of 40%, and coffee roasters 45%. The share of the retail price retained by coffee-producing countries declined from one-third in the early 1990s to only 10% in 2002, while the value of retail sales doubled.[10] The rapid concentration of retailing in Europe and the USA, where three or four firms often control 60% or more of food retail, has in its turn accelerated concentration in the final foods sector, with many market segments now being dominated by two or three firms. The rapid growth of urban food demand in developing countries is leading to this same model.

---

[9] Four firm concentration rates, i.e. the market share of the four largest companies.
[10] See World Bank (2007), *Focus D*.

In spite of these trends, SMEs have remained crucial in developed country food systems and are likely to be even more important in the developing country context given the importance of informal supply systems. SMEs are dominant in traditional agrifood activities, which escape the effects of scale and new quality demands. Local supplies are favoured where roads are inadequate, in areas of low population density and where modern distribution systems are now in place. SMEs are also emerging, however, in response to new market niches that demand innovation and entrepreneurialism. Greater outsourcing on the part of retail and food-processing firms is, similarly, opening up opportunities for small suppliers, often integrated into GVCs. More ambitiously, SMEs, organized in networks or clusters along the lines of the industrial districts of Italy, have been identified and are being promoted in many developing countries, often as a link in GVCs. And, finally, SMEs are emerging to the extent that new markets value products and production processes typically (although not exclusively) associated with family farming (Wilkinson, 2004).

Organics, fair trade and origin products

New high-value markets for food and other agricultural products that embody specific certified quality attributes, such as organics, fair trade and origin products, have increasingly become relevant in developed countries and some middle-income developing countries. With high rates of demand growth, these markets are regarded as potentially lucrative opportunities for exports of non-traditional products from developing countries (Henson, 2006).

Following Henson's (2006) survey, the world market for organic food and drink products in 2005 was estimated at US$24 billion, the EU accounting for 52% and the USA for 42%, together corresponding to almost 95% of global sales, roughly 40% of which was imported. Organic production in the EU accounts for 4% of total European farmland and is heavily subsidized (60% of organic land is included in policy support programmes). Around 0.2% of total farmland in the USA is under organic production. According to Dimitri and Oberholtzer (*USDA Amber Waves*, 2006), despite larger organic product sales in the EU, annual per capita retail sales in 2003 were nearly equal for both regions, approximately US$34 in the EU and US$36 in the USA. The sector's current growth rate is estimated at between 8% and 12% per year in Europe and 14–20% in the USA, a slower pace than that found during the 1990s, when it was around 20–30%. There is evidence that the rate of market expansion is slowing down further and that markets may become saturated in the near future (Henson, 2006).

Certified fair trade products have their origin in the developing world and demand reveals the interest of consumers in developed countries in the conditions under which agricultural products are produced and reach the retail market. Under fair trade, certified producers may receive premium prices for their products when compared to conventional producers. For instance, in Mali producers have received 70% more from selling cotton under fair trade rules and in Senegal they received 40% more (FLO, 2007). According to the World Bank (2007), however, recent studies confirm that the costs and margins for coffee sold through fair trade are high and that intermediaries receive the larger share of the price premium. One estimate is that farmers receive only 43% of the price premium paid by the consumer for fair trade roasted coffee and 42% for soluble coffee. The higher cost of processing and marketing would be partly explained by the diseconomies of scale related to small volumes and high costs, mostly associated with certification of supply chain actors, membership fees, advertising and campaigning.

Estimated retail sales of fair trade products in 2006 were roughly €1.6 billion, 42% higher than in 2005, but still an insignificant share of global food sales. Coffee and cocoa, traditional export commodities, have experienced the highest sale growth rates, at around 53% and 93%, respectively, between 2005 and 2006. According to FLO (2007), fair trade standards exist for food products such as tea, coffee, cocoa, honey, juices, wine grapes, fresh and dried fruits and vegetables, nuts and spices and non-food products such as flowers, plants and seed cotton.

Both organics and fair trade products have been identified as new export opportunities for developing countries. Nevertheless, as a corollary of the emergence of large middle-class markets in developing countries, both organic and fair trade are now also directing their attention to the domestic market. These categories of products can now be found not only in specialized shops, but also in the modern retail sector, which is now dominant in much of Latin America and expanding rapidly in Asia and a few countries of Africa (Raynolds *et al.*, 2007).

In a similar vein, 'origin-based' products are being promoted in developing countries. These often have their inspiration in the denominated origin products associated with Europe, now incorporated within the framework of TRIPS.[11] In the developing country context, however, many new features are incorporated – indigenous products, non-food products and products associated with the values of sustainability. The territorial association of origin products may also be more fluid and often on a much larger scale. To the extent that origin products are seen to offer market potential, they interest all classes of producer and in the developing country context the extent to which this niche will provide differential opportunities for small producers is not clear. CIRAD has been particularly active in the promotion of origin products in developing countries (Van de Kop, *et al.*, 2007)

Trade costs and the elasticity of substitution

We have called attention to the importance of FDI rather than trade, by highlighting the need for final food product industries to be close to the consumer market. Distorted tariff regimes, quotas and other trade barriers have also been identified as factors limiting trade flows. Here we complement this picture with two other features that impact trade and may reinforce the bias towards production close to the market.

First, evidence on trade costs breaks down into two major categories: costs imposed by policy (tariffs, quotas and the like) and costs imposed by the environment, such as transportation, insurance against various hazards and time

[11] Trade-related Aspects of Intellectual Property Rights.

Table 9. Commodity distribution of freight rates (as percentage of imports, aggregated over all partners). (From Hummels, 1999.)

	Average freight rate						
	USA	Argentina	Brazil	Chile	Paraguay	Uruguay	New Zealand
Food and live animals	14.1	21.6	23.1	21.9	12.8	7.4	15.4
Live animals	21.1	37.7	40.5	42.1	10.9	18.2	21.8
Meat and meat products	9.7	13.6	17.8	21.9	12.4	5.3	15.7
Dairy products	9.9	16.8	15.7	17.8	12	7.1	12.6
Fish	13.2	23.6	20.3	24.6	13	8.3	14
Cereals	14	23.3	27.9	27.9	13.4	9.1	20
Vegetables and fruits	17.4	23.7	24.7	21.7	12.6	7	16.9
Sugars	12.7	20.6	21.7	18.6	13.1	8.7	14.1
Coffee, tea	10.8	18	19.8	20.2	12.5	5.2	12.1
Beverages and tobacco	14.4	20.6	18.3	18.2	12.8	8	14
Crude materials	15.1	20.5	20.1	23.1	12	8.3	20.6
Minerals, fuels, lubricants	15.7	20.5	20.3	24.9	13.8	9.2	18.7
Animal and vegetable oils, fats	10.6	17.4	17.6	16.6	11.7	5	12
Chemicals	9	12.3	14	14.4	12.4	4.7	13
Manufacturing goods	10.3	15.5	17.7	14.7	13.3	7.9	13.1
Machinery and transport equipment	5.7	11.2	11.5	11.3	13.5	7.3	9.6

costs. There is another category of trade costs, associated with information barriers and contract enforcement, although this is not directly measured. More specifically, we focus on transport costs, which include freight charges and insurance, which is customarily included in the freight charge. Indirect transport costs include holding costs for the goods in transit, inventory cost due to compensating the variability of delivery dates and preparation costs associated with shipment size (Anderson and van Wincoop, 2004).

Hummels (1999) provides empirical evidence on the level and variation of freight and tariff rates at disaggregated commodity levels, including for imports into the USA, New Zealand and five Latin American countries (Argentina, Brazil, Chile, Paraguay and Uruguay) in 1994.[12] According to the author, clear patterns emerge. First, freight rates are lower for manufactured goods than for commodities. Rates are higher for agro-industry products, particularly for fruits and vegetables, cereals and crude fertilizers, among others. Second, landlocked Paraguay stands out as having exceptionally high freight rates; this may be an extremely important feature for LICs in sub-Saharan Africa and Central Asia. Illustrating these patterns, Table 9 displays ad-valorem freight rates, calculated as unweighted average rates of all observations within a two-digit SITC commodity group.

[12] In each case, data on import values, quantities (weights) and freight and insurance charges for each entering shipment are collected by customs officials.

Another fundamental feature impacting trade is the elasticity of import demand with respect to price and relative to the overall domestic consumption basket. Trefler and Lai (1999) present estimates over 1972–1992 for 28 industrial sectors in 36 countries. According to the authors' estimation, the agro-industrial sector generally presents low elasticities, such as 2.5 for food products, 1.9 for tobacco, 3.5 for textiles, 3.7 for beverages and 2.9 for wood products – all estimates much lower than the industrial average. Other interesting facts regard the negative correlation between the elasticity of import demand and import tariffs, as well as a significant negative correlation between the elasticity and the degree of market competition, usually implying that high elasticities are associated with markets dominated by less perfectly competitive industries.

Obstfeld and Rogoff (2000) argue that trade costs, including border costs such as tariff and non-tariff barriers, need not to be implausibly large to generate observed home bias consumption. According to them, such a bias may be generated by the interaction between preferences, trade costs and the elasticity of substitution between domestic and foreign goods.[13] Consumption home bias in agro-industry products may, therefore, be mainly promoted by relatively higher tariff and non-tariff costs, which would compensate lower elasticities of substitution.

Access to credit and industrial sector development

Rajan and Zingales (1998) ask whether industrial sectors that are relatively more in need of credit and external funding develop disproportionately faster in countries with more developed financial markets. They find this hypothesis to be true in a large sample of countries over the 1980s. The empirical strategy consisted of identifying an industry's need for external finance (defined as the difference between investments and cash generated from operations) from data on US firms. Under the assumption that capital markets in the USA are relatively frictionless, this method allowed the identification of the industry's technological demand for credit. Under the further and stronger assumption that such technological demand structure carries over to other countries, they examined whether industries that are more dependent on external financing grow relatively faster in countries that *a priori* were more financially developed.

For instance, the authors argue that, in countries that are less financially developed, an industrial sector such as drugs and pharmaceuticals, which heavily requires external funding, should grow relatively more slowly than the tobacco sector, which requires little external funding. In Malaysia, a relatively advanced country in terms of financial markets, drugs and pharmaceuticals had grown at a 4% higher annual real rate than tobacco. In Chile, which was in the sample's lowest quartile of financial development, drugs grew at a

---

[13] A suggestive rule of thumb for calculating home bias consumption derived from a microeconomic model consists of the formula $(1-\tau)^{1-\theta}$, where τ represents the level of trade costs, and θ the elasticity of substitution. For instance, with $\tau = 25\%$ and $\theta = 6$, the ratio of home expenditures on imports relative to domestic goods would be roughly 4.2, or, equivalently, the home bias would be around 80%.

2.5% lower rate than tobacco. Their estimates also show that, in general, financial development has almost twice the economic effect on the increase of the number of establishments as it has on the growth of the average size of establishments.

According to the authors' estimates on the patterns of external financing across industries in the USA during the 1980s, the agro-processing sub-sectors are among the industries that least require external funding for growth, particularly when compared to sectors such as electric machinery, radio, office and computers, drugs and pharmaceuticals. This information, although dated, brings added support for the relevance of the agro-industry sector for economic growth and employment opportunities in those developing countries generally associated with low levels of financial development. Agro-industry sub-sectors are relatively less dependent on external funding and access to credit, an argument that supports the findings on the sector's large contribution to total manufacturing value added, employment and overall turnover in developing countries.

Major Drivers of Competitiveness

The revolution introduced by the concept of agribusiness or agro-industry, first in the USA in the 1960s (Davis and Goldberg, 1957) and then in Europe (Mollard, 1978) and Latin America (Vigorito, 1978) in the late 1970s involved a fundamental rejection and revision of the traditional three sectors approach (agriculture, industry and services) to economic analysis. Agricultural activities now occupied an intermediary space whose processes were dependent on industrial inputs and whose products were the objects of industrial refashioning. To the extent that agriculture was now seen through the lens of industry, however, the agricultural space was understood to be increasingly residual, controlled and undermined by technological advances. In this light, the persistence of the small farmer sector could be reinterpreted as a consequence of this undermining of the autonomy of the agricultural space, which made it uninteresting for capitalist investment except in exceptional circumstances. Within this framework, biotechnologies could be seen as the final nail in the coffin, since agricultural products would become increasingly independent of specific soil and weather conditions and in some cases could even be reproduced industrially via fermentation (Goodman et al., 1987).

While many elements of this analysis remain valid it was an excessively production/technology push vision, which did not take into account another revolution in process at the same time, at the level of demand.[14] Seen initially as an industrial strategy of differentiation and segmentation in response to the stagnation of commodity-based food demand, consumption came to assume

---

[14] The renewed dynamic of commodity markets in the light of strong, sustained economic growth in large developing countries represents the clearest continuation of this dynamic and not surprisingly it is in this context that the first generation of biotechnology products has found its place.

more complex value traits. Some of these emerging values corresponded to broad demographic trends (ageing of the population, changes in the organization of family life) or new institutional contexts (shift in the public–private balance in questions of health and focus on prevention rather than intervention). Others, however, were less predictable, such as the sustained opposition to the application of genetic engineering. Perhaps most surprising of all, agricultural and artisanal rather than agro-industrial products became the norm for food quality, albeit with increasing deference to 'industrial standards'. As a result, the values of space and place have redefined agriculture's relationship with industry (and services, as we shall see), and the latter's inputs and processes have now to enhance rather than annul the 'natural' values of the agricultural product.

FDI and the role of demand

Reforms in the regulatory environment for market access and investment, together with the revolution in communications and logistics, have transformed the above-mentioned values of space and place into new forms of competitive advantage for developing countries. Some new global social movements, e.g. Via Campesina, and agro-ecologists have seen this as an opportunity to recapture the autonomy of agriculture in an unorthodox return to the three sectors approach to economic life. If, however, we integrate the market implications of the new 'quality turn' into an extended version of the agro-industry conceptual framework, now incorporating services, the possibilities for harnessing this new competitive advantage to broader development strategies become evident.

Within this scenario developing countries have a dual advantage. Their agricultural resources are more versatile and productive for a whole range of highly valued products in the global market. At the same time, a combination of population growth and urbanization is already making their domestic markets the most dynamic poles of the global system and they should continue to be so for the next 2 or 3 decades at least. If these structural advantages are to be transformed into the basis of sustained development strategies, however, the complex processes of globalization and transnationalization will have to be negotiated: how to combine market access with the consolidation of a domestic agrifood base which builds on the resources of small farmers and small-scale urban food operators; how to attract FDI in ways which complement and promote, rather than 'crowd out' domestic agrifood system actors; how to ensure that the incoming FDI and domestic initiatives respect environmental sustainability and do not lead to a 'race to the bottom'; and how to ensure a growing share in value added in global chains (the 'upgrading' challenge) where value is increasingly concentrated at the point of consumption. This challenge becomes sharper to the extent that we are dealing with value added in the consumer service sector. Here, upgrading must go beyond the notion of incorporating postharvest activities and explore forms of equity participation as in some successful initiatives within the fair trade movement (Wilkinson, 2007).

South–South trade

The emergence of the South as a global consumption pole has led to two import-ant new developments. In the first place, there has been an increasing flow of South–South trade (over 50% of Brazil's agricultural trade is now with other developing countries). This has been accompanied by a growth in South–South FDI and, to a lesser extent, in South–North investment as developing countries' consumer and export potential is reflected in the emergence of Southern transnationals. The South–South axis has been promoted by the growth of the animal protein complex. Here we are dealing with a comparatively small number of MICs, who are increasingly challenging the long-time hegemony of the USA, some European countries and even Australia and New Zealand.

The components of the animal protein complex, it should be noted, are very different from those of the 'non-traditional' exports sector. Agriculturally, with the exception still of poultry and pigs, they are land extensive and capital intensive and industrially scale is a precondition of competitiveness. This is the world of the global commodity traders, which will find new forms of expansion, again in the developing world, with the emergence of global biofuels markets. Hogs, but especially poultry, are an exception within this complex to the extent that they have been the privileged basis of small or medium farmer contract arrangements with agribusiness. Poultry, in particular, has become the para-digm for discussions on contract forms of coordination. Research has diverged sharply on the benefits accruing to small farmers but it is notable that this sec-tor has emerged as an important component of the domestic agrifood system in many developing countries (Little and Watts, 1994; Eaton and Shepherd, 2001; Birthal *et al.*, 2005). Even if income returns are comparatively attractive the opportunities of upgrading for primary producers in this value chain are quite limited given the scale intensity of the slaughter and processing stages.[15] Nevertheless, domestic firms and cooperatives control many poultry sectors in developing countries. With the market reforms, however, developing country poultry production has shown itself to be vulnerable to imports, with the par-ticularity that these now often come from other developing countries. This problem has been particularly acute in some sub-Saharan African countries and points to likely tensions within the developing country bloc as South–South trade flows gain in importance.

The Non-traditional Sector and Interpretations: Global Value Chain and Discovery Costs

The non-traditional sector appears to be moving to a large commercial farm model with a small farmer fringe on a subcontracting basis or to supply the domestic market. To the extent that this segment assumes greater importance,

---

[15] Artisan options have opened up with the emergence of free range, organic niches but even here scale is beginning to impose itself.

the farm worker, particularly it would seem the female farm worker, replaces the integrated small farmer of the poultry model. Much research has identified precarious working conditions in the non-traditional sector. At the same time, attention has been drawn to the beneficial effects of new social and environmental standards governing access to developed country markets, leading to considerable on-the-job training and the implementation of at least domestic minimum wages and health and safety regulations. Adjustment to a model that tends to consolidate development poles based on entrepreneurial wage-based agriculture may prove difficult to assimilate in developing countries where the strength of small farmer social movements has led to an institutional framework favouring small farmer strategies.

The literature on the non-traditional sector is broadly divided into two approaches. The GVC analytical framework widely adopted in cooperation programmes and by many national governments in developing countries tends to see the issue in terms of supplier zones linking with a predefined demand from buyer-driven GVCs (Humphrey, 2005). The strategic question then becomes that of identifying the conditions under which 'upgrading' may occur, which in the agro-industrial context would mean at least exporting ready-to-eat produce. The upgrading in question, however, is reactive, responding to the opportunities opened up by the global chain leader. Without adopting a GVCs approach other analysts of non-traditional exports have also emphasized the learning opportunities that emerge in an export sector finely attuned to varying market demands (Athukorala and Sen, 1998). Other authors have tried to refine the GVC analysis to take into account the specific features of natural resource value chains and have incorporated a sectoral approach based on Pavitt's now classic typology (Pietrobello and Rabelotti, 2006).[16] The results here, which point to the key role of public sector research, highlight the importance of public policy for the success of upgrading.

A second approach has focused less on the training and management aspects of non-traditional exports and more on the entrepreneurial challenges. This research has been inspired by the Hausmann and Rodrik (2003) hypothesis of 'discovery costs' in a developing country context. The argument is that while the new product or process is generally already known in the developed country context, where it is often protected by patents, there are a great many uncertainties with regard to the conditions of production in developing countries and the associated 'discovery costs' must be borne without the benefit of protection from imitative competition. The correct balancing of individual and welfare benefits, it is argued, points to the need for a more active intervention of the State. Research in Latin America, which explored these hypotheses in a considerable number of cases in eight countries, questioned the centrality of 'production costs', but reinforced the importance of a range of information costs requiring the provision of public goods (Sánchez *et al.*, 2006). In addition,

---

[16] Rather than organizing industry into product groups, Pavitt's taxonomy identifies four categories of industry from the point of view of technical change and innovation – supply dominated, scale intensive, specialized suppliers and science-based firms.

the case studies repeatedly drew attention to the key role of entrepreneurial initiative in the context of considerable uncertainty and risk. Various niche markets have been developed independently of any pre-existing value chains, such as blueberries in Argentina, sturgeon/caviar in Uruguay, and flowers in both Colombia and Ecuador, which have depended on innovating entrepreneurs. Independent initiatives of this kind require considerable financial resources and varied social and business networks. They are not, therefore, the likely products of a small farmer cooperative. It may be, however, that such initiatives are less likely today as the globalization and transnationalization of retailing absorb an increasing range of market niches (Wilkinson, 2007).

For authors who focus on the centrality of non-traditional exports for growth strategies, the central issues become those of market access and 'compliance' costs. It should be recognized, however, that tariff escalation and tariff peaks are still major obstacles to strategies of upgrading. Nevertheless, standards have become the predominant mechanism for regulating minimum requirements in global quality markets. Both the nature of these markets (specific quality features) and their global character in a context of transition from national and regional to global regulatory regimes have led to a complex mix of private (individual and collective) and public standards. While there is a consensus on the need for standards, their definition and evolution often appear arbitrary, opening up the suspicion of new forms of protectionism. The costs of quality barriers have been widely disputed. Evidence suggests that the majority of freight rejections relate to old-style quality issues (sanitary in the US and basic residue levels in the EU). Furthermore, it has been shown that the costs of compliance, although onerous, have been more than compensated by the gains from subsequent access, as in the case of shrimp exports from Bangladesh (Jaffee and Henson, 2004). Nevertheless, commercial standards can be extremely harmful, particularly for small countries dependent on a reduced number of export items (Oyejide, 2000). Research has called for a greater participation in the definition of standards within the different global forums, although this has been questioned as being technically and financially unrealistic for most countries (Athukorala and Jayasuriya, 2003). Rather, it has been argued that developing countries should concentrate on the specific conditions for compliance and negotiate their implementation with the help of various international cooperation programmes dedicated to this goal.

A less sanguine interpretation of the increasing importance of processed food exports from developing countries has been developed by environmentally oriented research, which would see this tendency as part of a broader movement either to export 'dirty' industries to, or deplete the resources of, countries with less rigorous legislative and regulatory controls. The fishing industry has particularly come under attack. A *United Nations Environmental Agency Report* (UNEP, 2001) warned of the dangers of selling rights to fishing stocks under the pressure for short-term export earnings, particularly when developed countries are subsidizing their fishing vessels. Research and social movements have also drawn attention to the environmental impact of aquaculture, particularly acute in the case of shrimp production (Wilkinson, 2006).

Retail, FDI and market redesigning

As a counterpart to the research on non-traditional exports, other research programmes have focused on the transnationalization of the commercial circuits in developing countries, through the increasing presence of global retail companies in their domestic markets (Reardon *et al.*, 2003). This research has focused on the systemic impact of global retailing to the extent that it is geared not only to the middle classes and the metropolitan centres, but also to the broad mass of urban consumers. The global food industry leaders, such as Nestlé and Unilever, have recently adopted a similar orientation to the low-income consumers of developing countries, as these have increasingly adopted modern retail shopping habits. The thesis of this research on the transnationalization of retailing is that the same system of quality and logistical standards is now redefining the conditions of access to the domestic markets of developing countries. If, as we have seen, the small farmer has difficulty in integrating into non-traditional export chains, he now also faces the same problems in accessing domestic, urban markets. Retailers are tending to set up their own distribution centres based on selected suppliers. These new circuits, in their turn, have a knock-on effect, leading to the modernization of traditional wholesale and outdoor markets, closing the door on those who are unable to adapt.

The speed and extent of these changes in developing countries have been challenged (Humphrey, 2006). In a number of Latin American countries, various transnationals have been unable to consolidate their presence and domestic or regional companies have strengthened their position. It may be, of course, that local retailers increasingly adopt similar standards to the global players as a condition of continuing competitiveness. On the other hand, there is considerable heterogeneity with clear gradations in quality demand. Some authors speculate that as basic quality improves retailers may shift back to traditional supply systems. In some major developing countries, such as India, foreign retailers are only now being allowed to gain a foothold. In much of Africa, it is the informal sector that predominates even in the large cities (Sautier *et al.*, 2007). In Brazil, global food industry companies, facilitated by new logistical technology, are focusing more on small outlets in an attempt to counter the buying power of large retailers. In India, there are preemptive moves to strengthen traditional distribution networks prior to the entry of global retailers.

While special attention has rightly been given to the importance of retail FDI since it involves redesigning the organization of food systems as a whole in developing countries, incoming FDI into developing countries has been particularly pronounced in food-processing and services. We mentioned earlier that the food industry, to the extent that it deals in final food products, tends to situate itself close to the relevant consumer markets. Trade, therefore, is proportionately less important than FDI compared with other industrial sectors and investments are diffused in accordance with consumer concentration. Nestlé, Unilever and other global food companies are typically present in as many as 150 countries. During the 1980s and 1990s most developing countries adjusted their legislation and regulations to attract foreign investment, which it was thought would enhance their export competitiveness. While this has been the case with the grain traders in the Southern Cone[17] and also for the 'maquila'[18] industry in Mexico, food industry investment has generally been motivated by the dynamism of developing country domestic markets.[19] Green field investment has primarily been restricted to regions of new export growth, as in the case of the crushing oil seed industry in the Mercosur. The food industry, for its part, has preferred mergers and acquisitions in consolidated oligopoly markets, relying on financial clout and global brands. With the recent turn, however, to low-income consumers, foreign firms have begun to operate in more competitive markets.

Whether foreign firms rely on imports of raw materials, equipment or human capital seems to depend on the conditions prevailing in the host country. Often, initial import dependence is replaced by local supplies once technical conditions are met. The degree to which FDI promotes rather than 'crowds out' the growth of domestic firms would also seem to be quite variable. While not necessarily bringing state-of-the-art technology in either products or processes, the incoming flow of FDI has tended to coincide with increased productivity and product diversification. This may, however, as has been observed, be due more to the new conditions of global market competition rather than the presence of foreign companies in the host country. Large developing countries such as China and India have placed various forms of restriction on FDI. In India, it has been explicitly excluded from certain sectors including, until very recently, the retail sector. In China, there has been a requirement for FDI to associate with domestic firms. An interesting series of case studies in China has also documented the way domestic firms have adjusted to the presence of FDI, repositioning themselves competitively within the local and regional market on the basis of closer knowledge of local food preparation and eating practices. The cultural specificity of food practices would seem to be promoting proportionately greater location of R&D in developing countries when compared with FDI in other industrial sectors.

While in the 1980s and 1990s the priority of developing countries was to attract FDI at all costs, the concern now is increasingly with more nuanced policies directing FDI to previously defined priority sectors. This is in line with a broader appreciation of the role of public policies in promoting development. As we have indicated earlier, a novel feature of FDI is the increasing presence of investments from developing countries both in the South and the North and the emergence of Southern food transnationals. This tendency is likely to become more pronounced in the coming decades as a result of strong sustained growth in developing country domestic markets. This will be particularly the case in Asia where conditions governing the flows of FDI have in the past been more restrictive.

A new wave of FDI, primarily directed also to the South, is now being stimulated by the priority being given to the development of biofuels. This surge

[17] Generally defined as Southern Brazil, Argentina, Chile, Uruguay and Paraguay.
[18] An industry that assembles parts imported duty-free for re-export to the country of origin.
[19] Japan, where FDI has been largely motivated by a strategy of re-exporting to its own domestic market, would be an exception here.

promises to revert the century-long specialization of agriculture towards food production. There are some indications, particularly in the case of ethanol, that the general intention is to create a separate global biofuels market. Initially, however, there is likely to be greater instability in crop prices, which is already making itself evident, and the accompanying food inflation will have differential impacts on producers/consumers and importers/exporters. The reintegration of energy into the basic 'functions' of agriculture has been welcomed by some as providing the opportunity for a new model of decentralized growth for developing countries (Sachs, 2006). Others argue that this will lead to renewed emphasis on the South's role as commodity crop exporter for the northern markets, substituting the traditional tropical commodities now in crisis (Seedling, 2007).

While the global traders and agrochemicals firms are heavily involved in these biofuel investments, FDI here is notable for the presence of new sectors. These include the automobile and petrochemical industries and extend to investment funds and finance firms, with Merril Lynch, Stark, Goldman Sachs, Soros, Rabobank and Barclays all being directly involved in the construction of biofuels plants. In addition, and this may also represent a shift from the model of contract coordination to the older 'colonial' pattern, investments include the purchase of agricultural lands for the production of agro-energy. Southern FDI is also engaged, particularly in Africa, a key target of biofuels investors, where the presence of Brazilian, Indian and Chinese FDI is increasingly evident. Given the importance of biofuels investments and the dual function of many crops, both food and energy production must be integrated into agro-industrial development strategies.

Traditional exports, commodities and downgrading

This potential reversion for the South to a commodity vocation within the emerging global biofuels markets is ominous in the light of the prolonged crisis in tropical export agro-industrial commodities. In some instances there has been a recycling out of these crops into non-traditional exports, as in the case of Kenya and, certainly in terms of revenue, 'non-traditionals' as a whole now overshadow the returns from tea, cocoa, coffee and cotton. Nevertheless, non-traditional products are often located in different regions and different countries, generating income and employment for different actors. Non-traditional activities tend to be more specialized, differing from the tree crop export culture, which was developed in synergy with subsistence production and producing crops for local markets. While the traditional sector continues to involve much broader numbers of workers/producers and their families in the developing world (10 million in cocoa and more than 20 million in coffee) farmers have at least some protection in their other crops.

While non-traditional crops involve a valorization of the agricultural and rural phases of production with corresponding opportunities for negotiating, not always successfully, greater shares in value added, traditional crops have experienced an inverse tendency. Not only has there been deterioration in the value of the raw material, but the value added has been concentrated at the service end of the chain, as in the explosion of the coffee shop culture. Representatives of the sector claim that producer country shares in global value of the coffee chain have declined from some 30% to 10%.

The value chain literature identifies a process of 'trading down',[20] rather than 'upgrading'. Research has highlighted the continuing importance of global traders in the conduct of traditional commodity (and other) GVCs, leading them to posit the 'bipolar' coordination of GVCs in which economic power within the chain is negotiated between traders and confectioners. Since the end of commodity agreements producer countries have lost control over stocks and have suffered competition from new entrants (such as Vietnam in the case of coffee). At the same time, new blending techniques and/or new processing techniques have enabled the use of lower-quality raw materials in both cocoa and coffee. In sharp contrast to non-traditional products, therefore, these commodities have suffered from 'downgrading'.

The huge numbers of farming families who depend on coffee in many different developing countries and the lack of perspective for upgrading within the chain provoked the emergence of a 'new economic social movement', the Fair Trade movement, which frontally challenged the justice of existing patterns of remuneration within the chain. This movement has had a striking impact – with some 600,000 families now covered by Fair Trade contracts in coffee – and has been reinforced to the extent that corporate social responsibility has become generalized within business. As a result, most global players in the chain, after years of resistance, have now endorsed (for whatever motives) the principles of the movement. At the same time, within the market, Fair Trade occupies a quality niche. We will not dwell here on the tensions within the movement and different evaluations of its dynamic. The important new phenomenon, which the movement reflects, is that social justice is now beginning to be seen as a quality factor from the standpoint of consumption.

A similar tendency can be seen in the emergence of a complementary movement around geographical indications (GIs), which in the developing world context (as a result of incorporation into the TRIPS agreements), are currently being applied to the production of these commodities. Some coffee-producing regions in Brazil and Colombia already have this status, as has Roibus tea in South Africa and Darjeeling tea in India. Here, again, the notion of quality is associated with social and cultural values relating to collective local development. To the extent that it anchors these qualities also in characteristics of the product, which are seen to depend on the particularities of the locale, however, it approximates to the revalorization of the rural and agricultural products and processes referred to in the case of 'non-traditionals'. While GIs defend social and cultural objectives through the claim for unique quality status, Fair Trade's goal is for the universalization of new criteria of redistributive justice within traditional commodity chains. Both strategies involve a

---

[20] See Gibbon and Ponte, *Trading down: Africa, value chains and the global economy* (2005). This position is close to the 'immiserating growth' hypothesis of Bhagwati and more generally the approach adopted by ECLA (Prebisch), whose more recent formulation was the 'spurious growth' scenario.

de-commoditization of the chain, which now depends on new forms of coordination (certification, seals, auditing).

Cocoa is moving towards a similar situation, but on the basis of a different dynamic, which is analysed by Fold (2002). Cocoa depends on forest regions, which are becoming scarce with accelerating deforestation. A low-quality strategy is therefore no longer sustainable, since the fundamental problem is the threat of irreversible declining supplies. The need to promote small farmer commitment to the renovation of cocoa production has led to a joint endeavour by processors and chocolate manufacturers, called the International Cocoa Initiative. The visibility of this programme and the explicit commitment of industry mark a rupture with the anonymity of the commodity market. In this context, civic concerns over child labour and labour conditions more generally must be taken on board and the 'bipolar' leadership of the GVC has had to be amplified to include international NGOs who are now involved in the Initiative. Fold concludes his analysis by calling attention to the way this International Cocoa Initiative has had the unintended consequence of introducing modern contract relations into this traditional commodity export sector.

Expanding urban markets

While most of the literature on development strategies has concentrated on the new export market opportunities within the historic South/North axis, FDI and trade is increasingly geared to the potential of the expanding urban markets in the South. Various research programmes and cooperation activities have focused on the actual and potential role of these domestic markets as the key for employment and income generation along the expanding agrifood chains, as they adapt to new producer–consumer relations. Particularly in the case of horticulture, the urban domestic markets have been shown to be five times as important in the case of Latin America (Reardon and Berdegué, 2003), and similar estimates have been reported in the case of developing countries in Asia and Africa (Shepherd, 2007). The retail research, which we have discussed above, however, has also cautioned that the corollary of transnationalization in function of the domestic markets of developing countries is the progressive internalization of quality and logistical standards before applied only to exports. Deregulation and the integration of developing country markets into global trade and investments also mean that the growth of the domestic market does not automatically revert to benefits in employment and income generation for domestic food producers and processors. As we have seen, LDCs have become on average significant net food importers.

In many low-income agricultural LDCs, the urban food system exhibits a dual structure. On the one hand, we have the large-scale processing of, generally imported, grains, milk and other products, mostly located in the country's ports and capital cities. The bulk of urban consumption, however, depends on a myriad of informal distribution and processing chains, primarily drawing on each region's small farming sectors and their traditional foods. This informal sector can account for as much as 80% of urban food consumption. The importance of this phenomenon and the need for appropriate institutional adjustments has been recognized in a paper by the African desk of the World Bank, which warns against the generalized imposition of standards, which 'could marginalize the important informal food delivery system and put special burdens on small food enterprises' (Jaffee *et al.*, 2003, p. 24).

An illustrative research study carried out by CIRAD (Sautier *et al.*, 2007) captures well the importance of this informal sector and particularly the key role of women in food-processing and catering services. The authors argue that urbanization does not lead to a sharp change in the nature of the foods consumed although the conditions of their supply are radically changed. Extensive social and ethnic networks maintain links between urban and rural dwellers and become the vehicle for adapting local foods to the urban context. In this process, previously untraded products, such as the cassava couscous 'attiéké' – which was only known in specific regions of the Côte d'Ivoire and is now traded in many Central African countries – become transformed from subsistence crops into components of the new urban diet.

In Dakar, Senegal, the number of urban cereals mills increased by 63% between 1990 and 1997; 339 informal service mills were identified and it was estimated that these artisan mills accounted for over 90% of the urban millet market. The World Bank report, previously referred to, also makes a complementary evaluation of transformations in the urban food-processing sector in the wake of the winding down of marketing boards and their processing oligopolies. The report argues that 'small-scale grain, oilseed, bakery and other operations (are) taking market share from the formerly dominant players and the latter (are) having to innovate their product lines and backward and forward linkages to survive, let alone prosper' (Jaffee *et al.*, 2003, p. 14). The CIRAD study, for its part, reports a survey conducted in Garoua, a secondary city of some 230,000 inhabitants in the Cameroon: 'A total of 1,647 small and micro commercial agrifood enterprises were identified, consisting of 866 food-processing units and 781 food preparing units (catering and street foods).' They note that being labour-intensive and decentralized these activities are important sources of employment and income for the poor, particularly for women, who managed 82% of these activities.

Farm–agribusiness linkages

The adjustment of developing country farming systems to the opportunities and challenges of their urban markets has been a central concern of international cooperation programmes including the FAO initiative 'Strengthening farm–agribusiness linkages' carried out in the three continents (FAO, 2004b). Since the 1990s, in the wake of the dismantling of many forms of domestic State intervention, national cooperation, the UN system and NGO networks have all focused on the micro-determinants of adjusting to market demand. Business management training, market research, innovative micro-financing and organizational promotion have all been harnessed to the goal of enabling traditional farmers to benefit from the income and employment opportunities opened up by urban food demand. These are precisely the enabling measures needed to transform rather than marginalize the informal sector, which at present plays a strategic role in domestic food supply systems. Methodologies and training programmes have been developed on all these issues, together with the identification of numerous 'success stories'. Nevertheless, the balance sheet is mixed (Shepherd, 2007).

The strengthening of traditional practices may initially be decisive in increasing food and nutritional security, but their long-term sustainability will depend on the degree to which they can be transformed into market assets. Value-added strategies based on tradition and locality, however, are successful to the extent that they are understood to be, if not unique, at least special. For most producers and producer groups, therefore, a focus on the dynamic products of urban food demand and an adaptation to the new conditions of access (management, marketing, logistics, quality) will provide the best conditions for success in many developing countries, particularly those where small farmer systems predominate. Whatever their successes, however, the microtargeting and support of specific producer groups is only relevant from a strategic perspective to the extent that it has a much broader ripple effect. In addition, therefore, to the perennial problems of reduplication, capacity substitution and hidden or not so hidden subsidies, which are either disproportional to the results achieved or unsustainable once removed, it is not clear that such a project focus can, of itself, provide the launching pad for agrifood development strategies.

It is increasingly recognized that the provision of public goods, and a proactive role for governments, is a necessary complement to the micro-focus on producer groups. There are no substitutes for adequate communication and information services, technical assistance and appropriate physical infrastructure. On their own, however, these latter services can favour specific producer groups to the detriment of others and may transform land-use patterns, leading to the marginalization of traditional actors. Both types of policies, therefore, are necessary, but their balance depends on the specific situation in each country, region and locality. International cooperation, therefore, cannot substitute for 'embedded' policy-enacting capacity, which in many cases means a renewed focus on the rebuilding of government competences.

Conclusions and Future Perspectives

Throughout this chapter we have seen that there are strong structural factors – in relation to both global and domestic markets – favouring the promotion of agro-industry in developing countries. The focus on non-traditional export products tends to revalue the natural climatic advantages of tropical and semi-tropical countries. In addition, the combined pressures of civil society social movements and advances in corporate social responsibility standards may well counter the 'trading down' effects of cheap wages and slack environmental regulation. Although the perspectives of long-term buoyant domestic demand, in a context of overall growth, will certainly lead to more imports, they will also provide a sustained stimulus to the development of domestic supply systems.

These combined advantages of developing countries have, as we have seen, increased inflows of FDI, whose potential benefits need, however, to be guaranteed by judicious policy measures. The force of developing country agro-industry can be seen in its increasing participation in South–South trade and investment flows. New contract relations are replacing spot markets, not only in the case of non-traditional products, but also in traditional tropical commodity exports, thus providing a more favourable environment for negotiating minimum social, economic and environmental standards.

The key challenges lie in the negotiation of 'upgrading' when primary production is inserted into GVCs, in establishing a level playing field in services, training and infrastructure for generalized competitive participation and in promoting an environment favourable to innovative risk-taking. Modernizing strategies for competitiveness should, however, be tempered by the recognition that tradition and the world of artisanal production and style-of-life consumption have an impact on increasing demand.

A strong case has been put forward for the employment and income benefits accruing to non-traditional exports with their labour-intensive, postharvest activities heavily biased towards female labour. A similar argument has been advanced for the spillover effects for non-agricultural rural employment of integrating small farmer communities into domestic markets. We have also seen, however, that many markets based on new quality and logistical standards provide formidable entry barriers to small farming and have tended to strengthen larger commercial farms where the small-scale farmer is at best reinserted as a wage labourer. These trends vary considerably from country to country and region to region but point to the need both for renewed attention to the implications of wage-labour agro-industry for rural development and for a discussion on ways of conditioning new investments on the adoption of measures of social inclusion along the lines of the Brazilian biodiesel programme.[21]

In spite of the favourable structural factors discussed above, issues relating to energy, global warming, innovation and the institutional and regulatory context all impose high levels of uncertainty with regard to long-term trends. We have already discussed how one scenario for biofuels would involve an unprecedented expansion of large-scale, wage-labour farming on all three continents. On the other hand, the costs of petroleum-based inputs open up competitive niches for labour-intensive, low-input or organic farming systems.

Global warming would seem particularly to threaten developing countries, from both floods and drought. Such threats, however, are not reserved to developing countries. One consequence has been the renewed attention to the development of drought-resistant varieties in genetic research. It now seems likely that advances here will be more rapid than originally projected as greater

---

[21] Different from Brazil's ethanol programme, the biodiesel initiative is specifically geared to involve the family farm sector not only as suppliers of raw material, but also in the production of crude oil. To achieve this the programme is designed to promote the sources of oil, which are most appropriate to each regional bioma. In addition, to participate in the programme biodiesel firms must obtain a social certificate which guarantees that they are purchasing the percentage of raw material from the family farming sector specified for each region (Wilkinson and Herrera, 2008).

resources are dedicated to this research. Developing countries, particularly in Africa, may well be important beneficiaries of such research, which may also shift perspectives on the acceptability of biotechnology and genetic engineering.

While changes in the institutional and regulatory climate have been unacceptably slow from the developing country perspective, as the impasse in the Doha round has made clear, many in Europe and the USA are already adapting to a 'post-commodity' farming scenario based on high-value products. To the extent that this develops, aided also by parallel movements in favour of local produce and against 'food-miles', the developed countries may become stronger competitors in the 'non-traditional' areas, which have until now provided such favourable perspectives for developing countries. Such a consideration reinforces the importance of gearing agro-industry strategies to the domestic urban food markets of developing countries.

Throughout this chapter it has been argued that agro-industry is the decisive component of the food system, intermediating raw material production in the rural context and consumption in the urban milieu. In addition, we have seen that its income, employment and location effects transform agro-industry into a powerful vector of broader development strategies. Our first conclusion, therefore, would be that policies for agro-industry should occupy a central position in government strategies. In addition, we have shown that agro-industry is a complex phenomenon involving global and domestic supply chains increasingly governed by contract, varied production systems, with different income and employment consequences, regulatory systems, research and development, FDI and international negotiations on quality, access and subsidies. Developing countries themselves are extremely varied in their natural and human resources and increasingly heterogeneous in their levels of economic development, making policy prescription hazardous. Nevertheless, some general policy implications can be drawn for developing countries from the different issues analysed throughout this chapter.

The first conclusion is that government-level initiative must now be given special attention. Second, however, it is clear from all we have discussed that policy must be oriented to market sustainability even when the values being transacted are traditional practices. Third, markets themselves are the objects of economic, social and environmental negotiation and regulation involving both public and private actors. Fourth, agro-industrial policies should also be a component of social policies aimed at food and nutritional security.

Last but not least, this chapter gathers empirical information from a great number of data sets, academic articles and books, multilateral agencies, NGOs, international forums and other web sites. In many cases minimum methodological uniformity was only achieved to the detriment of the number of countries surveyed, restricting our samples to only a very few countries. Important platforms such as that of the ILO or even the *UNIDO Industrial Statistics Database* suffer from the limited number of observations and generally do not allow for consistent inferences regarding absolute levels of the economic aggregates. There is evident scope here for cooperation among multilateral agencies since in many circumstances the same statistics appear for different years in different data sets, with different methodological approaches and, consequently, distinct results.

With these provisos in mind this chapter has highlighted the following areas as the privileged focus of policy initiatives:

- strategic policy on agro-industrial competitiveness;
- support for SMEs through capacity building, clustering and technology transfer;
- recognition of the key role of the informal sector and the need for appropriate enabling instruments; proactive policies in relation to FDI;
- policies for inclusion of small-scale farmers and agro-producers in contract supply chains;
- provision of public goods with a view to levelling the competitive playing field; participation in development of technical and monitoring services for achieving market access;
- provision of services for building up capabilities for sustainable market access; development of consumer protection policies;
- active role in harmonizing and ensuring the transparency of quality standards; measures to ensure that agro-industrial development is compatible with environmental and social sustainability to avoid 'the race to the bottom' trap as well as negotiation of standards and conditions of access in international forums.

Many of these policies will be best developed within the framework of concrete and effective international cooperation.

References

AEAA. 2002. *Statistics.* Agricultural Ethics Assurance Association of Zimbabwe, Harare, Mimeo.

Aksoy, M.A. 2005. The evolution of agricultural trade flows. In: Aksoy, M.A. and J. Beghin (eds) *Global agricultural trade and developing countries.* World Bank, Washington, DC.

Anderson, J.E. and E. van Wincoop. 2004. Trade costs. *Journal of Economic Literature* 42 (3) (Sept. 2004): 691–751.

Asea, P.A. and D. Kaija. 2000. *Impact of the flower industry in Uganda.* ILO Working Paper 148. ILO, Geneva.

Athukorala, P.-C. and K. Sen. 1998. Processed food exports from developing countries: patterns and determinants. *Food Policy* 23 (1): 41–54.

Athukorala, P.-C. and S. Jayasuriya. 2003. Food safety issues and WTO rules: a developing country perspective. ACIAR, Melbourne. Processed.

Barrientos, S. 1997. The hidden ingredient: female labour in Chilean fruit exports. *Bulletin of Latin American Research* 16 (1): 71–81.

Barrientos, S., C. Dolan and A. Tallontire. 2001. *Gender and ethical trade: a mapping of the issues in African horticulture.* Working Paper 26. Natural Resources Institute, Chatham Maritime, Kent.

Barrón, A. 1999. Mexican women on the move: migrant workers in Mexico and Canada. In: Barndt, D. (ed.) *Women working the NAFTA food chain: women, food and globalization.* Second Story Press, Toronto, pp. 113–126.

Birthal. P.S., P.K. Joshi and A. Gulati. 2005. *Vertical coordination in high-value food commodities: implications for smallholders,* MTID Discussion Paper No. 85, IFPRI, Washington, DC.

Blowfield, M., A. Malins and C. Dolan. 1998. *Kenya Flower Council: support to enhance-*

ment of social and environmental prac-
tices. Report of the Design Mission. Natural
Resources Institute, Chatham Maritime,
Kent, UK.

CIAA. 2006. *Data and trends of the European
food and drink industry*. CIAA, Brussels.

Collins, J. 2000. Tracing social relations in
commodity chains. In: Haugerud, A., M.P.
Stone and P.D Little. *Commodities and
globalization*. Rowman & Littlefield,
Lanham, Maryland.

Davis, J. and R. Goldberg. 1957. *A concept
of agribusiness*. Alpine Press, Harvard
University, Boston, Massachusetts.

Dijkstra, T. 2001. *Export diversification in
Uganda: developments in non-traditional
agricultural exports*. ASC Working Paper
47. African Studies Centre, Leiden.

Dimitri, C and L. Oberholtzer. 2006. EU and
US organic markets face strong demand
under different policies. *USDA Amber
Waves*, Feb. 2006.

Dolan, C. and J. Humphrey. 2000. Governance
and trade in fresh vegetables: the impact of
UK supermarkets on the African horticulture industry. *Journal of Development
Studies* 37 (2): 147–176.

Dolan, C and K. Sorby. 2003. *Gender and
employment in high-value agriculture
industries*. Agriculture & Rural Development Working Paper 7. World Bank,
Washington, DC.

Dolan, C. and K. Sutherland. 2002. *Gender
and employment in the Kenya horticul-
ture value chain*. Globalisation and Poverty
Working Paper. Overseas Development
Group, University of East Anglia, Norwich,
UK.

Eaton, C. and A.W. Shepherd. 2001. Contract
farming: partnerships for growth. *FAO
Agricultural Services Bulletin 145*. FAO,
Rome.

FAO. 2001. *Food balance sheets 1999–
2001*. FAO, Rome.

FAO. 2004a. *Statistical yearbook 2004*.
FAO, Rome.

FAO. 2004b. Strengthening farm–agribusiness
linkages in Africa. *AGSF Occasional Paper
6*. FAO, Rome.

FAO. 2007. *Challenges of agribusiness and
agro-industry development*. FAO, Rome.

FLO. 2007. *Shaping global partnerships*.
Fair Trade Labelling Organizations, Annual
Report 2006/07.

Fold, N. 2002. Lead firms and competition in
'bi-polar' commodity chains: grinders and
branders in the global cocoa-chocolate
industry. *Journal of Agrarian Change 2*
(2): 228–247.

Freeman, H. and J. Estrada-Valle. 2003.
Linking research and rural innovation to
sustainable development. Paper presented
at the 2nd Triennial Global Forum on
Agricultural Research (GFAR). May 22.
Dakar, Senegal.

Gelhar, M. and A. Regmi. (2005) Factors shap-
ing global food markets. In: Regmi, A. and
Gelhar, M. (eds) *New Directions in Global
Food Markets*. Agricultural Information
Bulletin No. 794, USAID.

Gibbon, P. and S. Ponte. 2005. *Trading down:
Africa, value chains and the global
economy*. Temple University Press,
Philadelphia, Pennsylvania.

Goodman, D., B. Sorj and J. Wilkinson, 1987.
From farming to biotechnology. Blackwell,
Oxford.

GRAIN. 2007. *Agrifuels*. Special Issue of
Seedling, available at www.grain.org/
seedling

Haggblade, S., P. Hazell and T. Reardon.
2005. The rural nonfarm economy: pathway out of poverty or pathway in? Paper
presented at the Future of Small Farms
Conference. June 25. Wye, UK.

Hausmann, R. and D. Rodrik. 2003. Economic
development as self-discovery. *Journal of
Development Economics* 72 (2), 603–633,
December.

Henson, S. 2006. New markets and their supporting institutions: opportunities and constraints for demand growth. Background
paper for the WDR 2008.

Hummels, D. 1999. *Have international transportation costs declined?* Working Paper,
Purdue University.

Humphrey. 2006. *Shaping value chains for
development*. GTZ, Eschborn.

Jaffee, S. and S. Henson. 2004. Standards and
agro-food exports from developing countries:
rebalancing the debates. *World Bank Policy
Research Working Paper 3348*, June.

Jaffee, S., R. Kopicki, P. Labaste and I. Christie. 2003. Modernising Africa's agro-food systems: analytical framework and implications for operations. Africa *Region Working Paper Series No. 44*. World Bank, Washington, DC.

Lawler, J. and V. Atmananda. 1999. *Gender and Agribusiness (GAP): case study of Cargill Sun Valley Thailand*. International Programs and Studies, University of Illinois.

Little, P.D. and M.J. Watts, 1994 (eds) *Living under contract: contract farming and agrarian transformation in sub-Saharan Africa*. University of Wisconsin, Madison, Wisconsin.

Mollard, A. 1978. *Paysans exploités*. Presses Universitaires, Grenoble.

Obstfeld, M. and K. Rogoff. 2000. The six major puzzles in international macroeconomics: is there a common cause? *NBER Macroeconomics Annual 2000*.

OCDE and FAO. 2007. *OCDE-FAO Agricultural Outlook 2007–16*. FAO, Rome and OECD, Paris.

Oyejide, T.A. 2000. *Trade policy and sustainable human development in Africa*. Africa Policy Dialogue, ICTSD, Windhoek, Namibia.

Palan, Z. and C. Palan. 1999. *Employment and working conditions in the Ecuadoran flower industry*. Sectoral Activities Programme Working Paper SAP 2.79/WP.138. ILO, Geneva.

Pietrobello, C. and R. Rabelotti, 2006. *Upgrading in clusters and value chains in Latin America: the role of politics*. BID, Washington, DC.

Rajan, R and L. Zingales. 1998. Financial dependence and growth. *The American Economic Review* 88 (3) (June 1998): 559–586.

Raynolds, L. 1998. Harnessing women's work: restructuring agricultural land and industrial labor forces in the Dominican Republic. *Economic Geography* 74 (2): 149–169.

Raynolds, L., D. Murray and J. Wilkinson. 2007. *Fair Trade: the challenges of transforming globalization*. Routledge, London.

Reardon, T. and J. Berdegué. 2003. The rapid rise of supermarket in Latin America. *Development policy review*. Blackwell, Oxford.

Reardon, T., C.P. Timmer, C.B. Barret and J. Berdegué. 2003. The rise of supermarkets in Africa, Asia and Latin America. *American Journal of Agricultural Economics* 85 (5): 1140–1146.

Sachs, I. (2006) *New opportunities for community driven rural development*, Instituto de Estudos Avançados USP, São Paulo www.iea.usp.br/english/articles.

Sánchez, G., R. Rozemberg, I. Butler and H. Ruffo. 2006 *The emergence of new successful export activities in Argentina: self discovery, knowledge niches or barriers to riches?* BID, Washington, DC.

Sautier, D., H. Vermeulen, M. Fok and E. Biénabe. 2007. *Case studies of agriprocessing and contract agriculture in Africa*, RIMISP, Santiago, Chile.

Senauer, B. and L. Venturini. 2004. The globalization of food systems: a conceptual framework and empirical patterns. University of Minnesota, Mimeo.

Shepherd, A. 2007. *Approaches to linking producers to markets*. FAO, Rome.

Trefler, D. and H. Lai. 1999. The gains from trade: standard errors with the CES monopolistic competition model. University of Toronto, Mimeo.

UNCTAD. 2005. *Trade and Development Report 2005*. UNCTAD, New York.

UNCTAD. 2006. *The Least Developed Countries Report 2006*. UNCTAD, New York.

UNEP. 2001. *United Nations Environmental Agency Report*. Geneva.

UNIDO. 2005. *UNIDO Industrial Statistics Database 2005*. UNIDO, Vienna.

Van de Kop, P., D. Sautier and A. Gertz (eds) 2007. *Origin-based products*. KIT-CIRAD, Amsterdam-Montpellier.

Vigorito, R. 1978. *Criterios metodologicos para el estudio de complejos agroindustriales*. ILET, Mexico.

Wilkinson, J. 2004. The food processing industry, globalization and developing countries. *Journal of Agricultural and Development Economics, eJADE* 1 (2): 184–201.

Wilkinson, J. 2006. Fish. A global value chain driven on to the rocks. *Sociologia Ruralis* 46 (2): 139–153.

Wilkinson, J. 2007. Opportunities for creating externalities in agroindustry and agrifood. CEPAL, Santiago, Mimeo.

Wilkinson J. and S. Herrera. 2008. *Making biofuels work for the poor – Brazilian case study,* OXFAM, Brasilia, Brazil.

Wilkinson, J. and R. Rocha. 2006. Agri-processing and developing countries. Background paper for the WDR 2008, RIMISP, Santiago, Chile.

World Bank. 2007. *World Development Report 2008: Agriculture for Development.* World Bank, Washington, DC.

WTO. 2007. *World Trade Report 2007.* WTO, Geneva.

WTO. 2008. *WTO Statistics Database.* WTO, Geneva.

4 Technologies Shaping the Future

Colin Dennis,[1] José Miguel Aguilera[2]
and Morton Satin[3]

[1]Director General, Campden BRI, Chipping Campden, Gloucestershire, UK;
[2]Professor, Department of Chemical Engineering, Universidad Católica de
Chile, Santiago, Chile; [3]Director of Technical and Regulatory Affairs, Salt
Institute, Alexandria, Virginia, USA

Background

We live in a time of great social, economic and technological changes. While
close to one billion people suffer from hunger or under-nutrition and another
two billion exist on the borderline of barely acceptable nutrition, the potential
for dramatically improving the economic status and food situation in developing countries has never been greater. Expected changes in income and demographics will lead to greater consumption of meat, dairy products, fruits,
vegetables and edible oils, resulting in a growing demand for raw agricultural
products. More consumers will have the economic status and changed lifestyle
which leads to the purchase of more processed and packaged food and an
increasing variety of convenience and luxury food items but does not necessarily increase the demand for raw agricultural commodities. The number of current low-income consumers lifted out of poverty will be the most important
determinant of the future global demand for food. The World Bank estimates
that the number of people in developing countries living in households with
incomes above US$16,000 per year will rise from 352 million in 2000 to 2.1
billion by 2030.

The ability of agricultural and food industries to continue to respond to the
undoubtedly substantial increase in demand in future decades will be highly
dependent on the increased application of existing technologies as well as the
exploitation of new and innovative technologies. By 2050, the world demand
for food will double, driven by the predicted population growth and the projected broad-based economic growth, which will lift low-income consumers out
of poverty.

The development of the agrifood industry will obviously vary in different
regions of the world depending on current levels of sophistication with
respect to the production, preservation and processing of agricultural commodities. Providing enough food for vulnerable groups of the population and

strengthening the competitiveness of the small farmers are probably the first priorities in developing and emerging countries. The focus will be on improving agricultural practices and on postharvest preservation technologies.

With changing demographic conditions and food demand there will also be increasing need for the design and development of efficient integrated systems of food production, processing, preservation and distribution from rural producers to expanding and diversifying urban populations in developing and emerging countries. In addition, the general shift towards increased meat consumption in developed and emerging countries is the greatest food transition of modern times. It is predicted that the world's livestock could eat as much grain as four billion people can by 2050 (Moynagh and Worseley, 2008).

In developed countries with well-developed urban populations there is continuing desire for greater added value and convenience of food production in response to the social and lifestyle changes (e.g. less time available for food preparation and greater disposable income) and an increased desire to consume foods which assist in disease prevention and healthy ageing.

The need for the exploitation of technologies is further emphasized by the fact that arable land and fresh water are not distributed around the world in the same proportions as is the population. For example, there are many barriers for Asia or the Middle East to be self-sufficient in food. With population growth, urbanization and broad-based economic development, food consumption in less-developed countries will outstrip their production capacity and will thus become larger net importers. This, in turn, will require appropriate transport and distribution systems.

It is well recognized that future food production will be constrained by land and water availability. There is, at most, 12% more arable land available that is not currently forested or subject to erosion or desertification. The area of land in farm production could only be expanded significantly by substantial destruction of forests and loss of wildlife habitat, biodiversity and carbon sequestration capacity. This is unacceptable from the viewpoint of environmental and natural resources protection. The only environmentally sustainable alternative is to at least double the productivity on the fertile, non-erodible soils already in crop production. In some areas land may be used in novel ways, for example, China's paddies produce two-thirds of the world's pond fish. During the 1990s, the country almost doubled yields per acre by growing multiple types of fish in the same pond (Moynagh and Worseley, 2008). In the near future, more fish may come from aquaculture than ocean fishing. It could be an effective way for people in poor countries to obtain the nutrients they desperately need.

Agriculture is not only the largest user of water (70% of fresh water) but also the largest waster of water. With rapid urbanization, cities are likely to outbid agriculture for available water. Thus, there will be a need for the world's farmers to face the challenge of doubling food production using less water than they are today.

Future strategies must not be restricted to ensuring food availability for all simply in terms of calories, but must also deliver sufficient quantities of safe wholesome food that contributes to a healthy diet. Strategies need to be considered in the context of progressive economic development and the associated urbanization. These will have consequences for the dietary patterns of lifestyles of individuals. Changes in diets, patterns of work and leisure are already contributing to the causal factors underlying non-communicable diseases, even in the poorer countries. Technological development therefore has a major role to play in shaping the future of food production, preservation and supply and delivery of food to the world's consumers.

This chapter considers the various drivers for technological change before discussing the range of technologies that will undoubtedly have a substantial impact on the development of the agro-food industry in developing, emerging and developed countries. These include specific processing and packaging technologies as well as the cross-cutting nature of generic technologies such as biotechnology, bioinformatics, nanotechnology and information and communication technology. Such technologies are discussed with respect to delivering health and well-being, ensuring food safety and contributing to more sustainable food supply in a competitive global market. Emphasis is given to the fact that technologies are not applied in isolation, but require commitment and investment from the private sector in a political environment where public policies stimulate entrepreneurship. This involves the availability of an appropriately educated and trained workforce, fiscal incentives for R&D and innovation and international regulations that are not unnecessary barriers to trade.

Drivers of Technological Change

Social

The attitude of consumers towards food and agriculture is heavily dependent on the availability and abundance of food in its various forms. In those parts of the world where the scarcity of food is such that individuals have only sufficient to satisfy their very basic calorific intake or are malnourished and suffer hunger, there is little thought given to the source of the food or its safety and quality. However, in those parts of the world where there is a plentiful supply, many consumers feel passionately about their food, its method of production, quality, origin and effect on their health, as well as its price. In no other industrial sector are there so many factors contributing to a direct consumer involvement in the products delivered. This provides both an enormous challenge and a huge responsibility for the agrifood industry.

In the past two decades or so most developed countries have seen a dramatic rise in concerns among their citizens over the quality and safety and long-term health effects of their food. A number of safety issues related to the food supply chain (local, national and international) have provided legitimate background for consumer groups to demand political action. For example, in Europe national food safety (food standards) agencies or authorities have been established, in addition to the European Food Safety Authority (Podger, 2005), to oversee the implementation of the regulatory framework, with a specific emphasis on safety in the broadest sense.

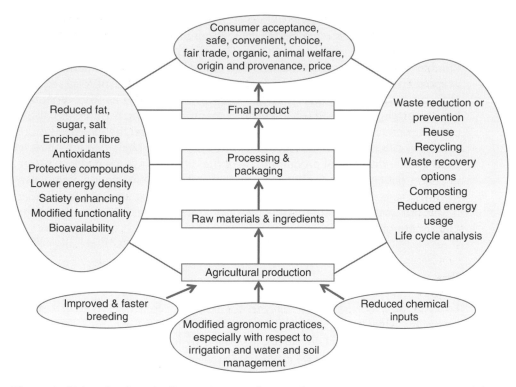

Figure 1. Future food production and processing trends.

Apart from safety, consumers are increasingly concerned about the origin of their food not only in terms of locality (region and country), but also about issues around animal welfare, environmental impact, organic production and fair trade (see Figure 1). Consumers increasingly have to make decisions about whether to purchase locally produced food versus imported products providing an all-year-round supply of, for example, fresh fruits and vegetables. The purchase of imported products often provides an opportunity for affluent consumers to support developing economies by purchasing their products. However, transporting food products over long distances (the 'food miles' debate) has stimulated much discussion on possible negative effects on the environment.

Consumers increasingly demand that food producers assure them that their ethical and environmental concerns are reflected in products. But, while all of the above factors play a part in putting pressures on the market for change, consumers remain very price-sensitive and seek solutions that are affordable. The need to embrace these diverse consumer concerns, yet provide foods that are affordable, places challenging constraints on the market and on the potential for innovation. Unlike all other categories of consumer products, where the consumer welcomes innovations and the application of scientific and technological developments, the outputs from science and technology with food products are often viewed with suspicion and the challenge to the industry is to communicate effectively the consumer benefits of scientific development.

In both developing and developed economies, consumers' increasing desire for choice, convenience and added value is continuing to influence the technological basis of the agrifood industry. This trend also includes the increasing number of meals consumed outside the home. In addition, the marked demographic changes (ageing populations) will influence the type of food required, the way it is packaged and the nutritional composition so as to contribute to healthy ageing. Increased urbanization in many developing economies will also pose challenges in relation to storage and distribution and the increased mobilization of different nationalities around the world will provide opportunities for even greater diversity of products to meet the different cultural needs. Changes in eating patterns from traditional diets to western-style foods and eating away from home (e.g. fast foods) will also be affected by rising incomes and the fact that more women are entering the workforce, leaving less time for food preparation at home.

Economic

Food security is not a new concern for countries that have battled political instability, droughts or wars. But for the first time since the early 1970s, when there were global food shortages, the issue of food availability is starting to concern more stable nations as well, especially as this will undoubtedly impact on food-price inflation (Anon., 2007f). There will be a permanent increase in demand for agricultural commodities in Asia, as the richer populations in China and India demand more protein. Demands for agricultural products from the biofuel industry, which is on course to consume about 30% of the US maize crop by 2010, will continue to have a major impact. These developments will underpin prices for the medium term. FAO estimates that these structural new trends will push the cost of agricultural commodities in the next decade between 20% and 50% above their last 10-year average. This will be a problem for economies where food represents a significant share of their import payments. FAO has forecast that lower-income 'food-deficit' countries will spend more than US$28 billion on importing cereals in 2009, double what they spent in 2002.

Several economic-related factors will influence both which technologies are applied in the future and where they are applied. Postharvest losses of foods – physical, nutritional and in market value – are inherent to a business dealing with perishable materials. Reducing postharvest losses by controlling temperature and moisture of stored grains, improved containers, packaging and cold chain maintenance of fish and horticultural perishables could, potentially, add to the global food supply and to the revenues of small farmers.

In many developing economies, there are tremendous opportunities for adding value in the country of origin of the raw materials. There is much greater opportunity for integration of the agrifood sector and the development of the organized food-processing sector in such countries. For example, new emerging economies such as in China, India and Brazil are seeing export growth of value-added products, while the value of exports as a percentage of the value of total world exports in food and drink products in both the USA and the European Union (EU) has declined by 30% and 15%, respectively (Anon., 2007a).

Key economic issues which will continue to affect the development of organized processed food industries are cost and availability of raw materials; labour availability and costs; rate of return on capital investment; transport costs and availability of distribution infrastructure; costs of obtaining regulatory approval for a new technology, ingredient or food; and cost of compliance with national and international regulatory frameworks.

Political

With the increased globalization of the agrifood industry, regulatory frameworks with respect to international trade have a fundamental part to play, especially with respect to food safety. The principle of equivalence in food safety has become an important issue impacting on international trade in foods (Anon., 2007c). At this level, considerable effort is being directed to achieving agreement in the concept of equivalence as it applies to food safety management systems, i.e. does the management of food safety in one country achieve or ensure the same level of protection as food safety management systems in a second country?

Equivalence of food safety measures is recognized in the World Trade Organization (WTO) *Agreement on the Application of Sanitary and Phytosanitary Measures* (SPS Agreement) and the *Agreement on Technical Barriers to Trade* (TBT Agreement). Both agreements require member countries to ensure that their food safety measures are objective, science-based, consistent and harmonized with international standards, where they exist. Because measures can take many forms, WTO member countries are encouraged to accept other countries' measures and regulations as being equivalent, provided they have satisfied these alternative measures and that the regulations meet their appropriate level of protection (ALOP) or public health goals. The ALOP, which is the responsibility of national legislators, may not be the same for all countries. The WTO has recently created the SPS Information Management System (SPS IMS), a database for searching for information on WTO member governments' sanitary and phytosanitary measures, which include food safety and animal and plant health and safety.

The Codex Alimentarius Commission (Codex), the international food standards setting body, is moving to better articulate the concept of equivalence and its application to food safety. The Codex Committee on Food Import and Export Inspection and Certification Systems (CCFICS) has developed guidelines for the judgement and development of equivalence of sanitary measures associated with food inspection and certification systems.

For food processors and food regulation authorities, there is also the need to determine the equivalence of different food safety measures, i.e. the ability of alternative technologies to achieve the same level of health protection by, for example, destroying or inhibiting pathogenic micro-organisms. The focus is on comparing existing approved measures, which are presumed to achieve a level of risk acceptable to the community, with alternative food safety measures.

Public policies with respect to food, diet and health will undoubtedly be a major driver for the agrifood industry in the future and will influence the need for technological development. Such policies could include intervention in relation to health claims in advertising, especially that targeted at children, and the way national governments respond to the need to resolve issues of obesity and being overweight, which is now occurring across all age and ethnic groups in both genders and across all socio-economic classes ('globesity') (Anon., 2007e).

Other political influences relate to public policies with respect to support of research and development and innovation or incentives to encourage industry to invest in new technologies. Similarly, policies with respect to encouraging education in science, technology and engineering will be a major influence on skilled and trained people essential for technological development. However, the commercial exploitation of new technologies will also depend on the culture of entrepreneurship and risk taking within a country or in individual companies and the public policies which encourage enterprise.

Environmental

There will be increasing pressures on the agrifood industry both from public policies and commercial need in relation to environmental issues throughout the food supply chain. These will include the need for lower and optimized use of fertilizers, pesticides, herbicides and fungicides according to weather conditions, growing season and soil types.

All the predictions on the consequences of climate change suggest that water availability is set to become a key issue around the world, with the associated major consequences for agricultural production and food processing. Such scarcity of water will strongly influence the use and methods of irrigation, plant breeding (e.g. drought resistance), water recycling and reuse in food production and processing systems.

Another environmental consideration that will influence development in the agrifood industry is that of waste (see Figure 1). A commonly adopted waste management hierarchy is waste reduction and prevention, reuse, recycling, other recovery options (including bioenergy) and, lastly, safe and environmentally sound disposal. All of these aspects will increasingly drive food production and processing systems in the future. The aim will be to develop and adopt production systems that are productive, sustainable and least burdensome on the environment. Thus, there will be increasing pressure to reduce emissions with respect to food processing and to decrease the carbon footprint of different systems. However, in order to target the appropriate part of the food supply chain and appropriate technologies, considerably more objective data from relevant life cycle analyses from farm to fork are required (Foster et al., 2006).

Technical and scientific

In addition to the above market-pull factors, technological development in the agrifood sector will also be shaped by current and future outputs from scientific and technological research and development.

For example, the desire to minimize the environmental impact of agriculture will focus attention on the potential benefits from greater application of integrated farm management, including emphasis on integrated nutrient management, which aims to increase the use of all nutrient sources (soil resources, mineral fertilizers, organic manures, recyclable wastes and biofertilizers). Similarly, decision support systems built around knowledge of the effect of agronomic conditions on plant growth and the onset and spread of pests and disease will be increasingly used together with satellite technology to optimize the application of fertilizers and pesticides or herbicides on specific crops.

Another major area of science that will drive technological development is that of nutrition. Research and development will continue to provide an improved understanding of the interaction between human psychology and physiology and food and drink. Important aspects include the following:

- Understanding of food structure and its influence on human physiology and nutrition. For example, it is now recognized that particle size, the structure of the food matrix and the proportion of amylose and amylopectin in foods can have a significant impact on blood glucose levels when food is ingested.
- Role of food constituents and food viscosity on energy intake. Greater knowledge on satiety may well offer the possibility of providing foods which assist in lowering energy intake and associated weight control.
- Production, formulation and separation of bioactive components and the effect of processing and delivery mechanisms on bioavailability as part of a normal diet.

Developments in material sciences will continue to enable the production of new materials for packaging, with the likely emphasis to be on biodegradable and compostable materials consistent with the sustainability agenda. Other developments are likely to focus on lighter weighting, recyclability and enhancements to consumer use, especially responding to changing demographics and meeting the needs of an ageing population.

Continued developments in automation and robotics will enable greater integration and automation of highly value-added, large-scale processing lines. Such developments will be enhanced by developments in vision and other non-invasive sensor systems which are integrated into feedback process control loops to ensure greater process reliability, product consistency and reduced waste or reworking of materials. Many such developments will be dependent on outputs from the basic sciences linked to the ability to store, mine and visualize large data sets.

As summarized in Figure 2, technologies that shape the future will have to contribute to safety and quality, especially in relation to nutrition, and

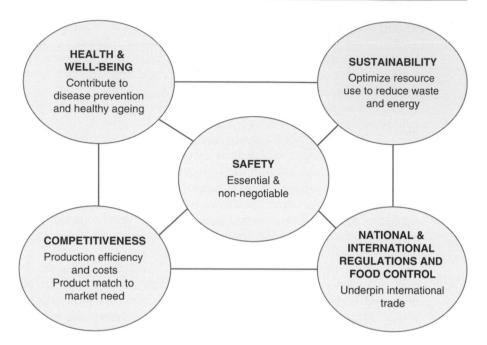

Figure 2. Drivers for technologies shaping the future.

sustainability (economic, social and environmental), while being competitive and complying with an international regulatory framework as part of increasing international trade.

Technologies for the Future

This section provides an overview of technologies that are likely to impact on a range of agro-industries in both developed and developing countries in the next 20–30 years, given the key drivers of the food industry and current global trends. Consumers' requirements largely condition the industry response in the use of technology. Improved convenience, higher quality and the demand for safer, healthier, fresher and more natural products have elicited a trend towards milder processing or combination of treatments, use of fewer additives and reduced packaging, among others. In addition, concerns about the environment and the use of energy are imposing new challenges to food-processing technologies.

Food-processing technologies

It is not easy to classify food technologies in a simple and succinct way that is at the same time technically rigorous. Figure 3 shows the scheme used in this chapter. There are, of course, other ways of classifying these technologies and of selecting them for given country-specific needs (van Boekel, 1998; Bruin

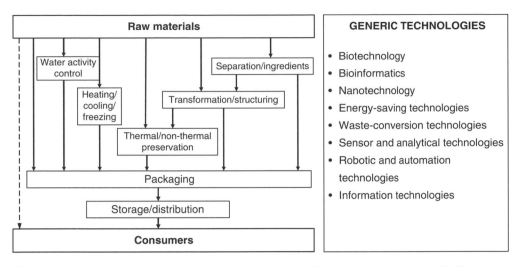

Figure 3. Scheme adopted to group technologies according to their main impact in the agrifood chain.

and Jongen, 2003). In developing countries many agricultural raw materials and fresh products are bought in nearby local markets and consumed at home without major processing as is the case of most fruits, vegetables, nuts and legumes and tubers (dotted line). Major staple foods that provide the bulk of calories in traditional diets of these countries are harvested, dried and stored, and undergo only cleaning and milling operations before consumption (e.g. rice, maize). Tuber and root staples, most notably potatoes and sweet potatoes, store well for extended periods and are peeled and cooked at home. Some components of crops are selectively fractionated and separated by industrial processing, becoming major ingredients of processed foods (e.g. wheat flour, oils and sugar) or high-value additives and flavourings. However, in industrialized societies and large urban centres in developing countries, most foods that reach the table have undergone some form of preservation to extend their shelf life and/or transformation to improve convenience and taste. The bulk of the processed foods industry involves fabricating foods by mixing, transformation and structuring technologies. Most foods experience some form of storage and packaging before distribution, which in advanced societies and large urban centres may be quite sophisticated.

For the three billion people presently living on less than US$2 per day, those technologies leading to increased agricultural output of staple foods, together with wider availability of storage facilities and improved postharvest practices, will contribute to their increased access to high-quality and safe food.

Annex 1, based on Figure 3, summarizes our views as to which technologies are likely to have a large impact in the agribusiness sector, with an emphasis on novel or emerging food technologies. As will be seen from Annex 1, many well-established technologies continue to undergo developments with the aim of improving product quality and processing and energy efficiency, while at the same time maintaining or improving the level of assurance of product safety. For example, in the traditional processing area of pasteurization and sterilization significant developments in the manufacture of expanded heat transfer surface per unit volume are occurring. One of the fundamental parts of a heat exchanger is the surface area for heat transfer. Significant advances are being achieved. Modern manufacturing techniques, such as direct laser deposition (DLD), allow complete freedom of 3D design and manufacture, with surface areas of $10,000\,m^2/m^3$ achievable (Schwendner *et al.*, 2001; Unocic and Dupont, 2003). New construction materials are being explored, such as polymer films instead of stainless steel. The result will be smaller heat exchangers for a given heat load, and at lower build costs. One of the first applications being investigated in the food industry is for recovering waste process energy from food factories.

Biotechnology

Experience to date suggests that biotechnology, if well managed, can be a major contributor to meeting future needs with respect to producing not only crops which are better adapted to a wider range of climatic and soil conditions (drought, salinity, acidity, extreme temperatures), but also crops that have traits for higher and better quality output (FAO, 2000). Modern biotechnology is not limited to the much publicized (and often controversial) activity of producing genetically modified organisms by genetic engineering, but encompasses activities such as tissue culture, marker-assisted selection (potentially extremely important for improving the efficiency of traditional breeding) and the more general areas of genomics, proteomics and metabolomics.

Second-generation genetically modified crops are expected to produce crops with higher levels of needed micronutrients, better quality proteins or crops with modified oils, fats and starches, to improve processing and digestibility. Developments will also undoubtedly occur which allow the production of specific functional foods or an enhanced level of bioactive compounds such as antioxidants.

However, the commercialization, promotion and diffusion of genetic modification will be tempered by concerns about the longer-term impacts and possible risks with respect to human health (toxicity, allergenicity) or for the environment (e.g. spread of pest resistance to weeds) and natural resources (modification of habitats). As indicated earlier the degree of caution any society will have about these developments depends on the societal preferences about their perceived risk and benefits (Thomson, 2002).

Bioinformatics

Bioinformatics is a powerful discipline that uses computing power to analyse biological data. As yet the full potential of bioinformatics has not been utilized by the agrifood sector; however, with a greater use of high-throughput technologies (microarrays, mass spectrometry) and the expansion of relevant

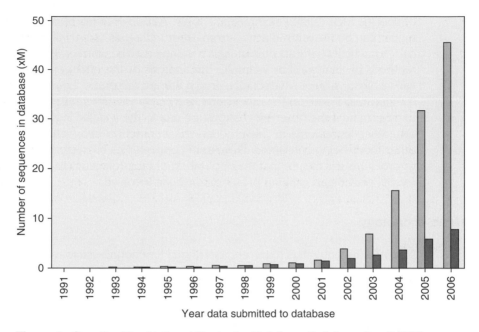

Figure 4. Growth of the National Center for Bioinformatic Information (NCBI) I database from 1991 to 2006. The total number of sequences (in millions) deposited in the GenBank (light grey) or protein (dark grey) databases is shown on the y-axis.

databases (see Annex 2 and Figure 4) this situation is likely to change. There are several areas where bioinformatics will prove invaluable to the food industry, including DNA and protein analysis for food authenticity, traceability and product development through the use of genetic markers (quantitative trait loci) in breeding programmes (Dooley, 2007). The newly emerging discipline of nutritional genomics, which uses many of the high-throughput techniques described above to improve the study of nutritional science and food technology, will also benefit from an increasing awareness and use of bioinformatics in the agrifood sector. Bioinformatics for protein analysis will be of benefit in terms of improving the understanding of protein properties during product manufacture; identifying proteins with specific functional properties, e.g. enzymatic function, identifying potential allergenic proteins or detecting potential bioactive peptides within protein breakdown products. Microbial analysis using bioinformatic tools will also be beneficial to the food industry for rapid pathogen identification and the development of beneficial microbial species for use in food manufacture. All these areas will benefit from increased speed, accuracy and automation brought about by bioinformatic tools that link laboratory-based technologies with analytical methods or reference databases. These will be of advantage to food producers, retailers, consumers and regulatory authorities, who all wish to ensure that high standards of product quality are maintained.

Further uptake of bioinformatics within the food industry is going to require an increased application of existing techniques along with the active development of specific food-related bioinformatic tools. Although many bioinformatic techniques can be transferred across from other industries, especially pharmaceuticals, food by its natural complexity is a unique matrix, which will require unique methods of analysis. For example, the analysis of the effect of a drug is basically a binary system whereby one drug is administered and the resulting change is measured. Food, on the other hand, is a composite material, so it is not easy to determine if the observed changes are due to the specific ingredient of interest, other ingredients in the product, the interaction between ingredients or other foods being consumed. Developing approaches to overcome these types of problems will ensure that the application of bioinformatics to the food industry will provide an exciting prospect for those involved.

Nanotechnology

Nanotechnology refers to the engineering of functional systems at the molecular scale. The prospect of the wide-scale use of products of evolutionary nanotechnology in food has engendered much debate. The concern is, if changing the size of materials can lead to radical, albeit useful properties, how size will affect other properties and, in particular, the potential toxicity of such materials. Although the products of nanotechnology intended for food consumption are likely to be classified as novel products and require testing and clearance, there are concerns, particularly in the areas of food contact materials, that there could be inadvertent release and ingestion of nanoparticles of undetermined toxicity. Such concerns need to be addressed, because the ultimate success of products based on nanotechnology will depend on consumer acceptance. The recent explosion in the general availability of products derived by nanotechnology makes it almost certain that nanotechnology will have both direct and indirect impacts on the agrifood industry (Anon., 2007d). Recent nano-based products include the following:

- Nanoparticles of carotenoids that can be dispersed in water, allowing them to be added to fruit drinks, providing improved bioavailability.
- A synthetic lycopene has been affirmed GRAS ('generally recognized as safe') under US FDA procedures.
- Nano-sized micellar systems containing canola oil that are claimed to provide delivery systems for a range of materials such as vitamins, minerals or phytochemicals.
- A wide range of nanoceutical products containing nanocages or nanoclusters that act as delivery vehicles, e.g. a chocolate drink claimed to be sufficiently sweet without added sugar or sweeteners.
- Nano-based mineral supplements, e.g. a Chinese Nanotea claimed to improve selenium uptake by one order of magnitude.
- Patented 'nanodrop' delivery systems, designed to administer encapsulated materials, such as vitamins, transmucosally, rather than through conventional delivery systems such as pills, liquids or capsules.
- An increasingly large number of mineral supplements such as nano-silver or nano-gold.

Potential future benefits from the application of the products of nanoscience and nanotechnology in the agrifood sector include application and effectiveness of agrochemicals, enhanced uptake and bioavailability of bioactive food ingredients, development of new tastes, flavours and textures and active and intelligent packaging, including new types of labelling, which aid traceability of products. There is also currently research on 'smart' surfaces that could, for example, detect bacterial contamination and react to combat infection. Although many of these materials contain nanoparticles, they are generally regarded as safe, provided their use does not lead to the release and injection of these particles. Concern has been expressed over the long-term fate and disposal of these materials, which might then lead to release of nanoparticles into the environment. These types of concerns will continue to stimulate debate on the labelling, approval, traceability and regulation of these nano-materials.

Food and packaging waste

With the increased emphasis on optimizing the use of natural resources and reducing or at least using waste, the EU has adopted a five-stage waste management hierarchy for use by industries in all EU states (see Figure 5).

Although waste reduction or prevention, reuse and recycling have been key aspects of cost reduction in manufacture for many years (Anon., 2006), there is now considerable interest in the possibility of creating energy from food and packaging waste (Anon., 2007b).

There are a range of technologies for the conversion of food waste to usable fuel or energy. The technologies differ in their stages of development, current commercial applicability, the scale at which they operate, the type of waste that can be processed and the form of energy produced. Although further developments are required, wider uptake in the food and drink industry would assist in reducing waste, increasing energy efficiency and contributing to future environmental and economic sustainability.

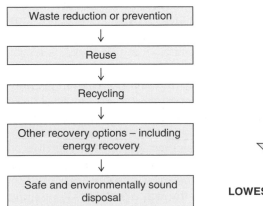

Figure 5. Waste management hierarchy.

Bioethanol production is currently a high-profile technology, particularly in the context of the 'food versus fuel' debate. The food industry has raised concerns that the growth of the bioethanol industry, and its use of energy crops, will have serious implications for the global food market as the two industries compete for the same commodities. This is especially true in countries where maize or cereals are used as feedstock. It would seem, therefore, that there is a need to divert the bioethanol industry away from the use of crops that could potentially be used for food, and towards the use of industrial waste materials as feedstock. For example, a Finnish energy company has established a pilot ethanol plant using waste produced on site at a Finnish food-processing company. Research should have the ultimate objective of widening the range of feedstock that can be used. Enzyme technology may be developed to improve the speed and efficiency of conversion of cellulosic wastes to a fermentable state, and genetic modification may result in the development of strains capable of yielding greater concentrations of ethanol in a shorter time than is currently achievable.

For biomass to fuel processes, the various challenges are the effect of moisture, waste types and composition and the inclusion of packaging materials on the efficiency of the process and the quality of fuel produced. A significant project is under way by a poultry processor in the USA to set up an on-site facility for converting animal by-product waste to synthetic crude oil. If successful, this technology could be applied to large meat and poultry processors elsewhere.

Anaerobic digestion is a relatively mature technology, for which the majority of fundamental research was carried out by the 1980s. Development work now focuses on areas such as effective pasteurization of digestates and the cleaning and upgrading of biogas. Novel reactor designs also allow scaling down and continuous running of the process. Design considerations should also take into account the inherent difficulties in controlling the anaerobic digestion process and sensors and monitoring systems developed that allow close control of feedstock processes according to the compositions of gases that are produced. Standard cultures for inoculating anaerobic digestion processes may be an area for research that would allow the processes to be better controlled. The use of manure, feeds and agricultural waste in biogas systems to produce electricity for village and small agro-industries in developing countries, through integrated rural bioenergy systems, is a promising development.

The conversion of waste oils and fats to biodiesel by transesterification is well developed and practised. Areas for research may, however, lie in the cleaning and treatment of both feedstock and the biodiesel product through filtration and dehydration.

Thermal techniques like gasification and pyrolysis produce fuels that are combusted soon after generation and the energy used as heat or for power generation. Incineration of biomass results in a high amount of heat that must be utilized immediately. The most effective way of utilizing the heat energy from thermal techniques is through a combined heat and power (CHP) system. The UK government has identified CHP as one of the best technologies to implement for the country to fulfil its commitments to greenhouse gas emissions under the Kyoto protocol. CHP increases the overall energy efficiency as it can co-generate both electrical power and heat energy. Energy efficiencies have been reported as high as 70–75% compared to the efficiency of sourcing heat and power separately, which are both around 30–40% efficient. Other co-generation systems, such as regeneration, are also gaining in popularity, as the drive for more efficient use of energy continues. Trigeneration systems are an extension of CHP processes as they provide the option of producing refrigeration using the heat in an absorption chilling process. This is particularly useful where refrigeration is a high operational priority and where excess heat may have no particular function and would otherwise go to waste.

While there is an awareness of technologies for converting waste to energy among relevant personnel in the food and drink industry, there is a general feeling that the technologies on the market are large-scale and unsuitable for the needs of individual companies. This is particularly serious for small and medium enterprises (SMEs) in developing countries. Small-scale or bespoke systems tend to be priced too highly and outweigh the benefits that they would offer in terms of energy and waste disposal savings. This scenario is likely to continue until the demand for such systems increases and brings down prices. Alternatively, there is the possibility of groups of neighbouring production sites collaborating on projects to establish centralized plants.

Waste-to-energy conversion systems and their operating parameters are generally tailored to the type and composition of the waste stream that they are designed to process. It is therefore advantageous if production processes continuously generate waste that is of a uniform composition. Processes that generate waste intermittently or that operate multi-product production lines may not realize the full benefits that a waste-to-energy system has the potential to offer.

Information technologies

In today's global economy, the ability to leverage *information* is critical to achieving competitiveness. The adoption of information and communication technologies is occurring at an incredible pace and will provide the core of potential for new entrepreneurs. For example, cell phones are now ubiquitous throughout Africa.

Access to these technologies removes constraining barriers between the entrepreneur and the marketplace. For the first time, the ability to connect directly to the markets allows the entrepreneur to achieve what had previously taken several intermediaries to deliver. At a very low cost an entrepreneur can set up a presentable web site that can influence buyers from around the globe. Of course, the entrepreneur has to be able to consistently deliver the quantity and quality of goods agreed upon in any contractual arrangement, but the fact is that there is now more direct contact between buyer and seller than ever before.

Organic coffee serves as an excellent example of the sort of niche markets that can profitably be accessed through the Internet. Through modern Internet auctions, farmers and processors have been able to achieve very significant sales volumes and prices. Previously, coffee inevitably changed hands many times between the producer and the buyer. In fact, most coffee sold today still moves under the old trader system. However, more and more coffee is moving directly. In 2007, the highest price paid for coffee was US$130 per pound for 100 lbs (US$13,000) of Panamanian coffee from a smallholder plantation through an Internet auction.

Potential for Technologies to Deliver Benefit in Different Development Scenarios

Health and well-being

Food technologies in years ahead will be increasingly targeted at providing health and well-being to consumers. To emphasize this trend, it is appropriate to add a new axis to the traditional food chain. When properly signalled by consumers' needs, the chain produces a flow of foods from the 'farm' to the table (mouth). Figure 6 captures this concept in a simple way.

This paradigmatic shift can be illustrated by reference to the flow of nutrients. Although the global average per capita consumption of food has risen by 17% over the last 30 years, to ~2800 kcal per day, the world still faces familiar problems of hunger and micronutrient deficiencies, but now accompanied by the ubiquitous presence of overweight people and obesity. This is also true for

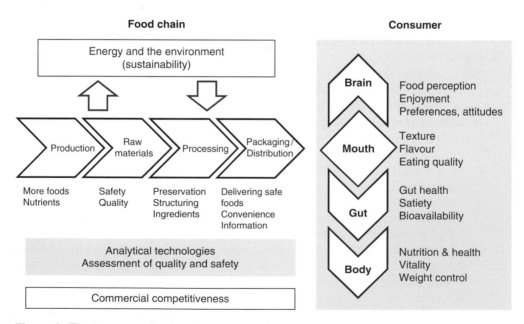

Figure 6. The two axes shaping the targets of food technologies for the next decades.

many developing countries. Increased food availability only makes sense when evaluated with respect to the impact on individuals, as it may have positive or negative effects. Moreover, there is increasing evidence that nutrients present in a food (i.e. as listed in food composition tables) may not be totally available for absorption in the gut. Also, absorption varies drastically (e.g. by up to 70%) for the same food depending, for example, on processing conditions and presence of other components in the diet. In many cases, processed foods show improved nutrient bioavailability when compared to raw or fresh foods that only suffer mastication prior to ingestion (Parada and Aguilera, 2007). The concept of nutrition and the impact of technologies may change, as we learn more about the fate of food components after ingestion ('food processing inside the consumer').

Food safety

Contaminated foods represent one of the most ubiquitous health problems in the world. They not only result in increased morbidity and mortality, but are also a major contributing factor to reduced economic productivity in many countries. The illnesses contracted from contaminated foods are generally caused by microorganisms (bacteria, viruses, moulds and their toxins), parasites, drug and pesticide residues, environmental pollutants (such as heavy metals, dioxine) and unconventional agents (e.g. bovine spongiform encephalopathy – BSE). They usually result in conditions such as diarrhoea, gastrointestinal pain, vomiting and headaches and, in the most serious cases, death. Worldwide, food pathogens have been estimated to cause 70% of the approximately 1.5 billion (10^9) cases of diarrhoea and three million deaths of children under the age of 3.

Because of their magnitude, very few countries have the ability and infrastructure to monitor the incidence of foodborne diseases. The Centers for Disease Control and Prevention in Atlanta (CDC) estimates that the number of cases of foodborne diseases in the USA is now equivalent to about 30% of the population per year. Although long-term chronic sequelae are characteristic of many foodborne diseases, they have not received the same degree of attention that the primary, acute symptoms have. Most foodborne disease victims and their physicians are happy to get over the short-term effects and seldom worry about future consequences.

Because of the difficulty in calculating the impact of long-term sequelae upon the health care system, they are seldom considered when determining the full impact of foodborne diseases. Long-term chronic effects are also overlooked simply because the data on them are not systematically collected and, as a result, it is difficult to link them directly to an originating cause. In fact, the significance of long-term sequelae is only beginning to be considered more comprehensively and the conclusions indicate that they can be very serious. Foodborne diseases have been implicated in many subsequent health disorders (Bula *et al.*, 1995; Smith, 1995; Stanley, 1996). Long-term chronic sequelae can destroy an individual's morale and quality of life and often result in measurable changes in personality.

In developing countries food safety has some specific implications beyond those listed above. Populations in these countries are particularly at risk because they do not have adequate supplies of safe water, appropriate waste disposal systems and access to refrigeration. Their low incomes preclude paying the extra cost involved in reducing food safety risks and countries may have only limited capacity to control the safety of foods. It is quite obvious that most efforts should be directed at avoiding the entrance of contamination sources into the food chain. In the second place, small-scale food industries and house-holds should be advised on the critical points involving food safety risks during manufacturing or food preparation at home. Even a traditional technology like fermentation that normally contributes to food safety in developing countries where refrigeration is not available needs the implementation of a hazard anal-ysis and critical control points (HACCP) system (Motarjemi, 2002).

There are some technological options that reduce the risk of contaminated foods. Irradiation is particularly suited to inactivate pathogens and parasites from fresh and dry foods, but its use depends on consumers' perception of irradiated foods. Other technologies are actively being explored for food decon-tamination, among them ozone treatment, pulsed and UV light and the use of electrolysed water. Physical, chemical or microbiological technologies can be used to detoxify grain and oilseed cakes by destroying, modifying or absorbing the mycotoxins so as to reduce or eliminate their toxic effects.

The application of appropriate technologies must be accompanied by good management and good hygiene practice along the supply chain as pre-requisites to an appropriate HACCP and traceability system. These, together with training and education of personnel, provide the essential components of a food safety management system (see Figure 7). The extent of such a system and its degree of complexity will depend on the size and complexity of the operation in question. However, the basic principles must be applied in all sizes of operation, whether relating to food retailing or the food service or catering sector.

Increased concern about food safety will affect developing countries in two major aspects. First, their exports will be exposed to increasingly demanding food safety standards from Codex Alimentarius and by unilateral requests from individual importers. Second, attitudes and standards in vogue in the developed world will spill over to the local market (Pinstrup-Andersen, 2000). A new form of protectionism may arise in which the high quality and safety standards imposed by importing countries may not be accommodated rapidly by local production technologies or guaranteed by local analytical capabilities, leading to an increased level of rejections at the entry ports.

Developing countries exporting fish and shellfish, as well as fresh fruits and vegetables, are likely to experience a more stringent inspection at the entry ports due to negative episodes involving food safety. For example, consumers in the USA have expressed their concerns over the fact that currently the FDA only inspects 1.2% of all imported seafood, which means that large quantities of contaminated products may be reaching the supermarket. Moreover, even if the problem regarding the safety of an imported food has been overcome, the credibility of the exporting country to produce safe food may be at stake, thus

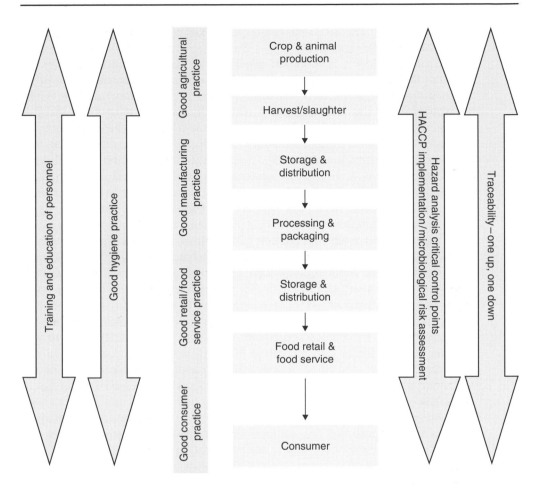

Figure 7. Integrated approach to food safety.

affecting the volume of its food exports. For this reason alone, developing countries should consider implementing or strengthening their foodborne disease control, investigation and surveillance systems.

Sustainability

Historically, the concern for the environment was not a matter of great appeal in the pursuit of economic gain. Throughout the centuries, large areas of the globe were laid waste as a result of mismanagement of resources. During the latter half of the 20th century, it became increasingly apparent that the earth was rapidly running out of the ability to support its burgeoning population. Following the UN World Commission on Environment and Development in 1987, the concept of sustainable development, the employment of socio-ecological processes that match the realization of human needs while at the same time preserving the quality of the global environment, became a generally accepted understanding.

The current status of economic development among countries was not the result of this view of sustainability. Natural resources, often usurped from other less powerful countries, were callously exploited and the environment boldly compromised in the rush to develop wealth. In some parts of the world this abuse is still taking place. Indeed, it is difficult for those that have not yet achieved an adequate level of economic development to accept the notion that global concerns for the environment will have to be respected, even though they may serve as a short-term constraint to growth. It makes it that much more challenging for developing countries to create an enabling environment and promulgate policies that will support the entrepreneurial sector.

The term sustainability also reflects a business concept related to the maintenance of a competitive position. The phrase 'Sustainable Competitive Advantage' or SCA was coined in 1985 by Michael Porter (Porter, 1985) and addressed as a potential strategy for the agro-industrial sector in developing countries development by FAO in 1986. The goal of SCA is to have entrepreneurs develop unique value-creation strategies that separate them from the competition. When first discussed in the context of agro-industries for developing countries, a number of constraints were readily apparent. Government policies seldom introduced incentives for entrepreneurial creativity. On the contrary, most policies stifled creativity. Very little effort went into coordinating agricultural production with agro-industrial processing; indeed, in most countries, the Ministries of Agriculture and Industry were competing agencies with very little interaction. Agro-industrial processors had little access to markets and competitive raw material resources. Within most developing countries and generally within the United Nations agencies, the overriding attitude was that it was the goal of agro-industry to serve the farmers of the developing world, despite the fact that, as a separate sector in the developed world, agro-industry far outstripped agriculture in terms of economic output and employment.

Thus, two different concepts of sustainability are at work in the creation of policies in support of the agro-industrial sector in developing countries: one, which manages access to raw materials and imposes controls upon unsuitable processing practices, in order to maintain the integrity of the environment and protect natural resources; and another, which requires an understanding of the competitive nature of the agro-industrial sector and the unique role of the entrepreneur. New technologies and new trade paradigms have a tendency to level the playing field – after that it is up to the players. Rational policies, a consistent approach and an enabling environment are what entrepreneurs require to compete successfully and be sustainable.

Competitiveness

Much has changed in the last 15 years. The globalization of the economy and the creation of the WTO have subjected small entrepreneurs to vastly greater markets that are becoming more visible and accessible. The wealthy developed markets are becoming more aware of processed goods available from developing countries. To supplement this, the growing sensitivity of consumers is motivating them to purchase products they perceive to be compatible with the classical concepts of social responsibility and environmental sustainability. This enhanced form of consumption is a large and growing sector. It includes high-quality commodities, such as selected or organic produce, value-added convenience foods, fair trade and environmentally responsive (e.g. bird-sensitive, shade-grown) products. Technologies such as irradiation are permitting exotic fruits and vegetables to travel much greater distances to reach profitable markets than they have in the past.

In light of these new developments, the more traditional ideas of competitiveness, such as economies of scale, may have to be reconsidered. Our concept of the comparative ability of an entrepreneur to sell and supply goods or services to a given market must consider how rapidly these markets are changing as well as how crucial changes to technologies have flattened the world and levelled the playing field for entrepreneurs. This is particularly the case for niche markets such as for organic foods.

One of the best indicators of how successful organic foods have been is that they have gone way beyond foods for human consumption and can now be found everywhere – even in the pet food markets. These types of purchasing patterns are all based on the notion that people want to express their individuality and there is no better way to do this than to indulge their specific tastes in foods and beverages. People also want to live a longer and better life and are suspicious of factory foods that contain many additives, even if their governments vouch for the safety of these foods. Of course, this is bad news for large processors, because their business depends on cost-effectiveness of food production, which, in turn, is the result of efficiencies obtained through large volume production. Large volume production, on the other hand, is not very suitable for an individual's specific preferences, which, of course, are better served by small operators. Historically, France and Italy have always been known as countries where you can find the best foods. It is not by chance that large multinational food companies are not nearly as popular on those markets as the thousands of domestic smaller manufacturers.

Twenty years ago, the number of different producers of juices and wines in the USA was quite limited. These days, that landscape has changed dramatically. Literally dozens of different organic juice manufacturers (including several local ones) are selling to supermarkets, and hundreds of new small-scale wineries are serving the desires of consumers – not to mention the multiplicity of imported wines from small foreign operations.

Our current understanding of nutrition recommends that we consume between 5 and 10 servings of fruits and vegetables per day. Although, there is usually a variety of fruits and vegetables available from local agriculture, consumers are broadening their horizons in the search for alternatives to conventional fruits such as apples, pears, peaches and bananas. The phenomenal growth in the market for mangoes is an excellent example of this. It is very likely that we will see the same thing in markets for many other tropical fruits.

And tropical fruits are not limited to the fresh varieties. Preserved fruits, by canning, by infusion with sugar or by drying, are also niche markets to look forward to. These products do not necessarily go only to individual consumers, but also to small manufacturers who want to include them as exotic ingredients in their baked goods or mixed cereal products. These changes are providing small-scale entrepreneurs in developing countries an ability to compete with far greater effectiveness than they have ever experienced before.

Technologies fostering agro-industrial development

The importance of technologies discussed in Section 3 is that they add value to raw materials or existing products. The value added may vary from incremental (e.g. a better package) to a radical change in production technology (e.g. a nano-based product). This is relevant in the sense that the impact of technologies should not be judged by the sophistication involved but by its relevance to better match the needs imposed by the final markets. Based on targets depicted in Figure 6, some major technology-based events that are likely to have a future impact on policies that foster agro-industrial development are presented in Table 1.

Conditions for successful adoption of food technologies

Technologies are not applied *in vacuo*; they are implemented by private entrepreneurs that sense a stable and propitious environment for their long-term investments. Under favourable conditions, the time frame in which the impact of food technologies can be fully realized may be only 10–15 years. For example, the Chilean salmon industry evolved from a quasi-artisan industry in the early 1990s to a world-class player in only 15 years; exports of farmed salmon in 2006 were US$ 2200 million. This figure represents one-half to one-third of the gross national income of three countries in the region. The role of governments in providing the human technical resources and the adequate regulatory framework must not be underestimated. There are other issues that also affect the adoption of technology, such as the support by equipment manufacturers and the availability of local technical personnel to implement and run the technology. In low-income and emerging countries the government can play a crucial role in providing support for new adopters of a technology.

In general, food policies as applied to technology tend to provide an enabling environment for food-processing entrepreneurs, create fiscal incentives for innovation, supply the necessary infrastructure for entrepreneurship and promote the adequate backward (e.g. financial support to SMEs, risk capital and information about future markets) and forward linkages (e.g. international promotion, 'country' brand).

The direct access to markets through information and communication technologies is probably the most important new development in recent history

Table 1. Technical implications for policy to foster agro-industrial development based on the identified technological trends.

Trends	Technical implications
Need for more foods, driven by rising incomes	Reduction of postharvest losses by improved storage and better marketing channels
	Adoption of processing technologies that foster the supply of processed raw materials
Demand for high-quality and safe foods	Adoption of novel technologies that preserve freshness and supply better taste and flavour
	Critical evaluation of emerging preservation technologies as to their equivalent effectiveness compared to proven technologies
Consumption of internationally traded foods	Development of appropriate traceability systems based on information technologies (ITs)
	Adoption of non-destructive inspection technologies for quality control
	Creation or strengthening of a regulatory framework attuned with international agencies
Foods for health and well-being	Design foods for the gut (e.g. functional foods) and the brain (gastronomy)
	Select processing technologies that preserve nutrients, secure functionality and provide a high bioavailability
Increased markets for organic products	Adoption of organic production systems and presence of reliable certification organizations
	Adapt preservation processes and packages that are non-invasive and replace synthetic additives by natural ones
Exports of value-added products	Develop human resources, technical infrastructure and technology transfer capabilities
	Build infrastructure and distribution chains for refrigerated and frozen products
	Cater to niches that require specific processed products (fresh and dried exotic fruits, etc.); strengthen quality management capacity
Environmental concerns	Strengthening of integrated management systems
	Adoption of life cycle assessment as evaluation criterion of impact of processing technologies
Globalization of market information by the Internet	Widen access to wireless communication technologies in rural areas and improve command of foreign languages at school level
Knowledge-based food industries and biorefineries	Strengthen the Science and Technology base at universities and national research institutes
	Apply advances in biotechnology and keep abreast of developments in nanotechnologies

for entrepreneurs. However, the potential to capitalize on this will largely be dependent upon the economic policies that are or are not in place. Government policies may be based upon a number of realities and business principles:

- No technology (processing or information/communications) will flourish in an economic environment not prepared to support it.
- It is the entrepreneur who takes the greatest risk and should receive the greatest reward if profits are achieved.
- It is a very competitive world and, in anticipation of an initial, non-profitable period, policy support has to endure a certain amount of 'staying' power (often up to 5 years) until profits start to flow. Grace periods in loans, for instance, can be made to address this issue.
- The entrepreneur's chief task is to make a profit – everything else is secondary to that goal.
- Policies in support of successful entrepreneurs will have the consequent benefit of generating employment and collecting taxes that can be put to social use.

Policy makers in developing countries interested in fostering entrance into food export markets should be aware that competing in these markets requires the use of production technologies that are of the same standard as those issued in the receiving (usually high-income) countries. To become and stay as major players in export markets, technological policies should address also the following complementary issues:

- Availability of well-trained local technical personnel with command of internationally spoken languages to run the production and processing aspects, as well as marketing operations. This imposes added competences to be built through the educational (technical and university levels) and continuing education system.
- A basic science and technology and innovation system that provides support to the local industry and promotes the entrance of new small and medium entrepreneurs into the business. This local talent is at present most probably located in universities and government research institutes. Specialized centres for adaptation, demonstration and transfer of technologies in areas with validated market potential will have to be implemented and sustained through time for the support of small and medium agro-industries (as is now the case with agricultural research units).
- Geographical associations in the form of interconnected technology clusters where suppliers, food processors, government agencies and institutions such as universities, research centres and trade associations merge to empower the innovation process.
- A basic infrastructure for roads, ports and connectivity (communications) that links producers and consumers inside the country and across the globe.
- A central regulatory food authority that protects consumers' interests locally and abroad and assures that food produced and exported meets the highest standards of food safety and hygiene.

Just as agro-industrial entrepreneurs are forced to operate in a competitive environment, so must farmers and all actors along food chains be forced to do so. Farmers have to be prepared to provide agro-industry with the right product at the right time at the right price. In order for farmers to get a better understanding of the risks and benefits of today's access to expanded markets, cooperative and partnership systems and arrangements between themselves and processors should be encouraged. This will allow for a more coordinated supply chain and will greatly increase overall competitiveness.

Because of the competitive nature of international trade, it is critical that governments take part in and send their most qualified people as negotiators to international fora such as the Codex Alimentarius Commission. Although these meetings generally revolve around technical matters such as standards and analysis, because of the litigious nature of the proceedings and the implications for fairer trade, consideration must be given to representation by highly trained negotiators and well-briefed legal people who understand the long-term significance of trade standards for their country. The subject of these meetings may be technical, but the consequences are definitely economic.

At the present time, the opportunities for developing country entrepreneurs to effectively compete in international agro-industrial trade are greater than ever. The spirit of entrepreneurship is alive and vibrant in developing countries. The use of modern scientific and information technologies accompanied by the supportive policies and instruments that will create an enabling environment will allow this spirit to flourish and benefit everyone in the country.

Conclusions

As observed in the initial section of this chapter, changes are taking place in the nature of food demand and in the socio-demographics that drive them. To face this changing world, we have developed an extraordinary set of new technologies that have the potential to produce food when it has never been produced before and in greater quantities than hitherto imagined. With little vision, one can picture the genetic manipulation of plants in order to grow them on previously non-arable land and under hydroponic conditions with quantum increase in quality and output. Likewise, our output of meat and fish has increased dramatically due to new management systems. Global warming must be monitored with extreme precision and is dramatically injuring many countries. International information exchange has developed to an extent that we can truly say we live in a global village. Political divisions will become less and less significant, as people from all parts of the world become empowered with the ability to communicate directly with one another. We are on the cusp of a revolution in the global movement of goods and services that would have been impossible to imagine a decade ago.

With reference to the movement of food, technologies and systems with the ability to support extended distribution chains will become increasingly important. Ready access to foods from around the world will have consequences for the dietary patterns of all individuals. As we have seen, together with the movement of goods and people, there is a distinct possibility of creating pandemics through the parallel movement of infectious diseases. This makes the promulgation and implementation of internationally harmonized high-quality standards imperative, such as those of the Codex Alimentarius.

For those countries whose economies allow consumers to think beyond the cost of food, they often incorporate social, ethical and environmental dimensions into their choices. These supra-economic dimensions of food may vary in different countries. While increasing global interactions may, in time, bring a certain degree of harmonization to these dimensions, they do provide an opportunity for producers and processors to fill specific market niches.

As more and more developed country economies move from a manufacturing to a service economy, food processing will relocate to transition and rapidly emerging developing countries. Food-processing entrepreneurs in developing countries will gain greater access to both niche and mainstream markets. In both cases this will require the growth of a regulatory and distribution environment that can deliver product of the desired quality, on time and at the right price. This will require considerable investment for the development of production and distribution infrastructure along with the training of technicians, managers and regulators.

The international movement of goods necessitates significantly greater attention to food safety. While the SPS and TBT Agreements are predicated upon science-based international standards, individual countries may opt for differing levels of protection depending on their particular requirements. However, because so many goods will be entering the flows of international trade, harmonization of science-based safety standards is likely to take place. This will have a profound impact upon food production and processing policies and practices, as well as on the technical and managerial training required to carry them out.

Future food production will face increasing challenges from a number of seemingly contradictory imperatives. The first is the need to produce more food with assured safety and increasing consumer appeal. However, this has to be accomplished in an atmosphere of increasing responsibility for maintaining the environment for future generations. All production must be sustainable, so the former freedom to use pesticides and fertilizers with little concern for the environment has been significantly curtailed. On top of that there are growing markets for products that no longer reflect the efficiencies of mass production. Thus organic, bird-safe, dolphin-safe and fair trade products are making major gains in high-end markets.

This will lead to a two-tiered food production system. It may allow smaller entrepreneurs, who do not have the capital to invest in large-scale production, to compete effectively on a smaller scale.

Regardless of the scale of agriculture employed, the environment will play an increasingly critical role in production. Global warming may result in changes to water availability, which will require adjustments in agricultural technology as well as broadening the scale of desalination to include impaired ground waters in addition to sea water. This latter technology will require the development of alternative energy sources, such as osmotic pressure power generation. The importance of agricultural waste management will favour the development of production systems that are least burdensome to the environment.

There is little doubt that biotechnology will become a major contributor to future production and processing technologies. The technology will not only focus on improving quantitative outputs, but will also be put to work to produce crops with higher levels of beneficial nutrients, such as antioxidants that are able to withstand longer distribution chains and harsh processing conditions.

Our knowledge of human genetics and nutrition continues to make fundamental contributions to improved health and disease prevention. The simultaneous analysis of genetic make-up and nutrient need will result in foods designed to meet a wider range of products focused upon health and well-being. Advances in the disciplines of genomics, proteomics, bioinformatics, nutritional dynamics and the nanosciences will be incorporated in foods that meet the individualized needs of people with specific genetic make-ups, occupational and lifestyle choices and stages in life. The technologies that will shape the future of the agriculture, fisheries and food sectors will provide greater safety, be more socially and environmentally responsible, provide the elements of better health and maintain a higher quality of the products' extended lifespan than any preceding goods.

Among the food technologies that are expected to play a major role in the future of food processing will be preservation techniques based upon sterilization and pasteurization; non-thermal technologies, such as irradiation and ultra-high-pressure processing; technologies that control water activity including microwave and freeze-drying, hurdle technologies and minimal processing; those based upon the extraction and isolation of specific food components, such as antioxidants, flavours, specialized lipids and other functional ingredients. Agricultural products may be bioengineered to produce large outputs of these specific materials and modern extractive technologies, such as supercritical extraction, will be employed to yield healthier, higher-quality products with a reduced negative impact on the environment.

Texture, mouthfeel and friability are among the sensations most evident to consumers. Optimizing these characteristics requires technologies designed to ensure specific food structures at all stages of production and in the finished product throughout its life cycle. Advancements in emulsification and gelation will utilize complex interactions of proteins, lipids, carbohydrates and water to develop flow, viscosity, tensile strength and plasticity to arrive at the most appealing textures.

Reduced barriers to trade have opened the way to much longer distribution chains requiring products that will maintain safe hygienic quality for longer periods and will meet all the sanitary and phytosanitary needs of importing countries. The traditional technologies of heat treatment will be supplemented by cold processing methods such as ultra-high-pressure processing and ionizing radiation, both of which are capable of producing products of the highest quality. These technologies will become increasingly important as the recognition for the need to significantly increase our consumption of fruits and vegetables becomes more apparent.

In order to meet future demands for longer shelf life, food will be maintained in optimum condition through developments in packaging materials and in modifications to the atmosphere immediately surrounding the products. While the technology itself is not new, there have been recent developments that have resulted in significant improvements in reducing microbial spoilage, as well as detrimental enzyme and chemical activities. Different combinations of oxygen, carbon dioxide, nitrogen and ethylene at differing levels of humidity are used to alter package atmospheres.

More recently, research has focused upon employing various types of packaging materials to be active, intelligent or interactive in atmosphere management. Such packaging can enhance the quality or safety and impart desirable characteristics to the food by altering atmosphere permeability through sensing and response to changes in the ambient environment.

As food products become increasingly focused on providing health and well-being, additional dimensions to the traditional food chain must be considered. As knowledge of the complex interactions of digestion develops, we will gain a more comprehensive view of the whole diet. The interactions between various nutrients, the role of various fibres in governing bioavailability and moderating water balance between the gut and the renal system, and the contribution of essential micronutrients by intestinal micro-organisms will all factor into future understanding of nutrition.

Product and technology selection will, to a large extent, be an extension of a country's agronomic potential coupled with the state of its economic development. For less-developed countries, the focus will remain upon establishing a functional and efficient food chain, largely to serve local and national needs. As development and potential increase, more advanced technologies and distribution systems will be employed.

In order to support the successful growth of food industries, policy makers will have to provide entrepreneurs with an enabling environment. Where possible, infrastructure development, encouragement of cross-chain business linkages and economic incentives should be established to stimulate growth. A food science and technology innovation system based upon a country's natural endowments and potential competitive advantage must be supported along with a regulatory food authority to protect the interests of consumers and to assist national entrepreneurs in gaining access to international markets.

Industrial development policy should not add to the risk of entrepreneurs, but encourage the application of sound, proven methods for the production of useful goods. It is important to consider sustainability in context and carefully adjust its significance within the hierarchy of imperatives weighted to achieve rational and successful industrial development. The globalization of the economy and the creation of the WTO have provided entrepreneurs vastly greater markets. In light of these new developments, policies must support entrepreneurial competitiveness in rapidly changing markets.

While we have seen the great cache of physical, chemical and biological technologies available to entrepreneurs for the production and processing of foods, perhaps the greatest tool available to them is the explosion of information and communication technologies. While food-processing technology has empowered entrepreneurs to produce products of higher quality, convenience and market potential, information and communication technology has provided the direct access and connections to promote and sell them.

For the first time, entrepreneurs in developing countries have a very strong potential to access international markets with an unprecedented degree of independence. However, the potential to capitalize on this will largely be contingent upon the economic policies that are in place. Such policies must provide strong support to the country's entrepreneurial base. When properly implemented, these policies will have the consequent benefit of generating employment and general economic development, which all will benefit from.

References

Ahvenainen, R. 2000. Minimal processing of fresh produce. In S.M. Alzamora, M.S. Tapia & A. Lopez-Malo (eds), *Minimally Processed Fruits and Vegetables*. Gaithersburg: Aspen, pp. 277–290.

Anon. 2006. Food Industry Sustainability Strategy. Available at: http://www.defra.gov.uk/farm/policy/sustain/fiss/pdf/fiss2006.pdf.

Anon. 2007a. CIAA Benchmarking Report 2007 Update. The competitiveness of the EU Food and Drink Industry. Available at: http://www.ciaa.eu/documents/brochures/Benchmarking_report_update_2007.pdf.

Anon. 2007b. *Conversion of Food Waste to Energy*. Food Processing KTN and Resource Efficiency KTN, UK.

Anon. 2007c. Equivalence in Food Safety Management. IUFoST Scientific Information Bulletin. Available at: http://www.iufost.org/reports_resources/bulletins/documents/IUF.SIB.Equivalence.pdf.

Anon. 2007d. Nanotechnology and Food. IUFoST Scientific Information Bulletin. Available at: http://www.iufost.org/reports_resources/bulletins/documents/IUF.SIB.Nanotechnology.pdf.

Anon. 2007e. Obesity. IUFoST Scientific Information Bulletin. Available at: http://www.iufost.org/docs/IUF.SIB.Obesityrev2.pdf.

Anon. 2007f. The end of cheap food. *The Economist*, December 8, pp. 11–12.

Artes, F., Allende, A. 2005. Minimal fresh processing of vegetables, fruits and juices. In D.-W. Sun (ed.), *Emerging Technologies for Food Processing*. Amsterdam: Elsevier Academic Press, pp. 677–716.

Augustin, M.A., Sanguansri, L. 2008. Encapsulation of bioactives. In J.M. Aguilera & P. Lillford (eds), *Food Materials Science: Principles and Applications*. New York: Springer, pp. 577–601.

Barbosa-Cánovas, G.V., Vega-Mercado, H. 1996. *Dehydration of Foods*. New York: Chapman & Hall.

Bourlieu, C., Guillard, V., Vallès-Pamiès, B., Gontard, N. 2008. Edible moisture barriers: materials, shaping techniques and promises in food product stabilization. In J.M. Aguilera & P. Lillford (eds), *Food Materials Science: Principles and Applications*. New York: Springer, pp. 547–575.

Bruin, S., Jongen, Th.R.G. 2003. Food process engineering: the last 25 years and challenges ahead. *Comprehensive Reviews in Food Science and Food Safety* 2, 42–81.

Brunner, G. 2005. Supercritical fluids: technology and application to food processing. *Journal of Food Engineering* 67, 21–33.

Bula, C.J., Bille, J., Glauser, M.P. 1995. An epidemic of food-borne listeriosis in western Switzerland: description of 57 cases involving adults. *Clinical Infectious Diseases* 20(1), 66–72.

Cen, H., He, Y. 2007. Theory and application of near infrared reflectance spectroscopy in determination of food quality. *Trends in Food Science and Technology* 18, 72–83.

Chua, K.J., Chou, S.K. 2005. New hybrid drying technologies. In D.-W. Sun (ed.), *Emerging Technologies for Food Processing*. Amsterdam: Elsevier Academic Press, pp. 535–551.

Clark, J.P. 2006. Pulsed electric field processing. *Food Technology* 60(1), 60–67.

Dooley, J. 2007. Bioinformatics: a review of current and future applications in the food industry. Review No. 59, Campden & Chorleywood Food Research Association, UK.

FAO. 2000. World Agriculture: Towards 2015/2030. An FAO Perspective. Available at: http://www.fao.org/DOCREP/005/Y4252E/y4252e03.htm.

Farkas, J. 2006. Irradiation for better foods. *Trends in Food Science and Technology* 17, 148–152.

Foster, C., Green, K., Bleda, M., Dewick, P., Evans, B., Flynn A., Mylan, J. 2006. Environmental Impacts of Food Production and Consumption: A Report to the Department for Environment Food and Rural Affairs. Manchester Business School. Defra, London. Available at: http://www.defra.gov.uk/science/project_data/DocumentLibrary/EV02007/EV02007_4601_FRP.pdf.

Gomez-López, M., Ragaert, P., Debevere, J., Devlieghere, F. 2007. Pulsed light for food decontamination. *Trends in Food Science and Technology* 18, 464–473.

Holdworth, D., Simpson, R. 2008. *Thermal Processing of Packaged Foods*. New York: Springer.

Holmgren, K. 2006. New technology expands probiotic applications. *Food Science and Technology* 20(4), 57–58.

James, C., James, S. 2006. Keeping it cold: storage and transportation. *Food Science and Technology* 20(4), 39–40.

Kulozik, U. 2008. Structuring dairy products by means of processing and matrix design. In J.M. Aguilera & P. Lillford (eds), *Food Materials Science: Principles and Applications*. New York: Springer, pp. 439–473.

Lee, J., Wang, X., Ruengruglikit, C., Gezgin, Z., Huang, Q. 2008. Nanotechnology in food materials research. In J.M. Aguilera & P. Lillford (eds), *Food Materials Science: Principles and Applications*. New York: Springer, pp. 123–144.

Leistner, L., Gould, G. 2002. *Hurdle Technologies: Combination Treatments for Food Stability, Safety and Quality*. New York: Kluwer.

Lopez-Rubio, A., Gavara, R., Lagaron, J.M. 2006. Bioactive packaging: turning foods into healthier foods through biomaterials. *Trends in Food Science and Technology* 17, 567–575.

Master, A.M., Krebbers, B., van der Berg, R.W., Bartels, P.V. 2004. Advantages of high pressure sterilization of food products. *Trends in Food Science and Technology* 15, 79–85.

Motarjemi, Y. 2002. Impact of small scale fermentation technology on food safety in developing countries. *International Journal of Food Microbiology* 75, 213–229.

Moynagh, M., Worseley, R. 2008. *Going Global, Key Questions for the 21st Century*. London: A&C Black Publishers Ltd, pp. 71–88.

Parada, J., Aguilera, J.M. 2007. Food microstructure affects the bioavailability of several nutrients. *Journal of Food Science* 72, R21–R32.

Pinstrup-Andersen, P. 2000. Food policy research for developing countries: emerging issues and unfinished business. *Food Policy* 25, 125–141.

Podger, G. 2005. Creating the European Food Safety Authority. Twenty-seventh Annual Campden Lecture. Available at: http://www.campden.co.uk/publ/pubfiles/lecture%202005.pdf.

Porter, Michael E. 1985. *Competitive Advantage: Creating and Sustaining Superior Performance*. New York: The Free Press.

Raoult-Wack, A.L. 1994. Recent advances in osmotic dehydration of foods. *Trends in Food Science and Technology* 5, 255–260.

Reglero, G., Senorans, F.J., Ibañez, E. 2005. Supercritical fluid extraction: an alternative to isolating natural food preservatives. In: G.V. Barbosa-Canovas *et al.* (eds), *Novel Food Processing Technologies*. New York: CRC Press, pp. 539–553.

Rico, A., Martín-Diana, A.B., Barat, J.M., Barry-Ryan, C. 2007. Extending and measuring the quality of fresh-cut fruit and vegetables: a review. *Trends in Food Science and Technology* 18, 373–386.

Sakai, N., Hanzawa, T. 1994. Applications and advances in far-infrared heating in

Japan. *Trends in Food Science and Technology* 5, 357–362.

Sanguansri, P., Augustin, M.A. 2006. Nanoscale materials development – a food industry perspective. *Trends in Food Science and Technology* 17, 547–556.

Sanz, P.D., Otero, L. 2005. High-pressure freezing. In D.-W. Sun (ed.), *Emerging Technologies for Food Processing*. Amsterdam: Elsevier Academic Press, pp. 627–652.

Schwendner K.I., Banerjee R., Collins P.C., Brice C.A., Fraser H.L.1. 2001. Direct laser deposition of alloys from elemental powder blends. *Scripta Materialia* 45(10), 1123–1129.

Scotter, C.N.G. 1997. Non-destructive spectroscopic techniques for the measurement of food quality. *Trends in Food Science and Technology* 8, 285–292.

Skurtys, O., Aguilera, J.M. 2008. Applications of microfluidic devices in food engineering. *Food Biophysics* 3, pp. 1–15.

Smith, J.L. 1995. Arthritis, Guillain-Barré syndrome and other sequelae of *Campylobacter jejuni* enteritis. *Journal of Food Protection* 58(10), 1153–1170.

Sorrentino, A., Gorrasi, G., Vittoria, V. 2007. Potential perspectives of bio-nanocomposites for food packaging applications. *Trends in Food Science and Technology* 18, 84–95.

SPS Information Management System. A database for searching for information on WTO member governments' sanitary and phytosanitary measures. Available at: http:// spsims.wto.org.

Stanley, D. 1996. Arthritis from foodborne bacteria? *Agricultural Research*, October, 16.

Strumillo, C. 2006. Perspectives on developments in drying. *Drying Technology* 24, 1059–1068.

Sun, D.-W. 2004. Computer vision – an objective, rapid and non-contact quality evaluation tool for the food industry. *Journal of Food Engineering* 61, 1–2.

Thomson, J.A. 2002. *Genes for Africa: Genetically Modified Crops in a Developing World*. Cape Town, South Africa: UCT Press.

Unocic, R.R., Dupont, J.N. 2003. Composition control in the direct laser-deposition process. *Metallurgical and Materials Transactions B* 34(4), 439–445(7).

van Boekel, M.A.J.S. 1998. Developments in technologies for food production. In W.M.F. Jongen & M.T.G. Meulenberg (eds), *Innovation of Food Production Systems*. Wageningen: Wageningen Pers, pp. 87–116.

Vermeiren, L., Devlieghere, F., van Beest, M., de Kruijf, N., Debevere, J. 1999. Developments in the active packaging of foods. *Trends in Food Science and Technology* 10, 77–86.

Wallin, P.J. 1997. Robotics in the food industry: an update. *Trends in Food Science and Technology* 8, 193–198.

Wang, L., Weller, C.L. 2006. Recent advances in extraction of nutraceuticals. *Trends in Food Science and Technology* 17, 300–312.

Yam, K.L., Takhistov, P.T., Miltz, J. 2005. Intelligent packaging: concepts and applications. *Journal of Food Science* 70: R1–R10.

Zhang, M., Tang, J., Mujumdar, A.S., Wang, S. 2006. Trends in microwave-related drying of fruits and vegetables. *Trends in Food Science and Technology* 17, 524–534.

Zheng, C., Sun, D.-W. 2004. Vacuum cooling for the food industry – a review of recent research advances. *Trends in Food Science and Technology* 15, 555–568.

Annex 1: Food Processing and Preservation Technologies Shaping the Future.

Preservation technologies. Food preservation technologies delivering products that are microbiologically safe and of high quality over their defined shelf life will continue to be of the utmost importance. Several non-thermal technologies hold this promise, particularly those in which preservation effects are delivered through 'transparent' packaging. To achieve superior products, reduced costs and increased energy efficiency, often more than one preservation method (e.g. combination technologies) will need to be applied. Environmental effects, convenience and safety issues will add increasing demands on packaged products.

Heat pasteurization and sterilization of foods

Technologies	Current status	Potential developments and factors
Thermal processing: Microbial load reduction is affected by thermal action (Holdworth and Simpson, 2008)	Canned foods enjoy an excellent record as affordable, convenient and safe. UHT-sterilized liquid foods delivered in laminated carton packaging are well established around the world	Canning continues to be exposed to public scrutiny due to energy and environmental considerations. Expansion of HTST alternatives for better quality and nutrient retention. Developments in retort systems to increase container agitation for increased heat transfer and reduced processing times
Retortable pouches: Use of flexible Al foil/plastic laminates instead of tin cans	Technology has been around for 4 decades as an alternative to canning with minor commercial impact. The reduced heating time due to the slender profile results in better-quality sterilized foods	New laminates, higher capacity production lines and improved seal reliability are required. Lighter weight should save energy during transport
Aseptic processing: Sterile product filled under aseptic conditions	Presently mostly limited to sterilization of liquid foods, including those containing food particles. Aseptic bulk storage and transport are important technologies in international trade	Extension to semi-solid foods by new heating technologies. Increased use of lighter weight and less-expensive packaging. Improved heat exchanger design to increase process efficiency and save energy

Non-thermal technologies for pasteurization and sterilization of foods

Technologies	Current status	Potential developments and factors
Ionizing radiation (IR): Effect caused by a highly penetrating form of energy that damages DNA of cells	Technology established and legal in many countries. Widely used for disinfection of dried produce (e.g. spices) and prevention of sprouting/disinfection. Recent applications include checking of emerging pathogens (Farkas, 2006). Limited acceptability in supermarket foods	Applications to eliminate infectious food pathogens and parasites and to extend the shelf life of many perishable foods will depend on risk assessment by regulatory agencies and consumer attitudes towards IR
Ultra-high pressure (UHP): Subjecting packaged foods to pressures up to 800 MPa	Texture, flavour and nutrient retention of UHP foods is normally better that those of thermally processed products. Used commercially in a few niches for pasteurization (juices, guacamole, oysters) due to high costs	UHP processing of a wider range of products will depend on the reduction of capital costs, and more efficient continuous systems. UHP sterilization is expected (Master et al., 2004). New packaging materials needed
Pulsed electric fields (PEF): High-voltage pulses induce membrane breakdown in cells	Demonstrated mainly to inactivate food-poisoning micro-organisms in liquid foods. Commercial application of PEF for juices with superior flavour quality reported recently (Clark, 2006)	Applied for pasteurization of liquid foods, providing adequate safety and improved quality
Pulsed light (PL): Use of short pulses rich in UV-C light	Lethal photochemical effects on micro-organisms known for decades. Only two companies are presently producing PL disinfection systems (Gomez-López et al., 2007)	Role in substitution of chemical disinfectants that may be harmful to humans and cause ecological problems. Used for surface decontamination, potential for application to packaging. Problems to be solved: heating spots and effects on nutrients
Biological preservation: Use of bacterial metabolites	Food-grade bacteriocins proven to inhibit many pathogenic and spoilage micro-organisms (e.g. in minimally processed foods)	New 'natural' bacteriocins are approved. Applications for short-term preservation (e.g. raw milk in less developing countries – LDCs), or pasteurization if followed by thermal or other non-thermal treatment

continued

Annex 1: Continued.

Technologies that control water activity. Control of the water activity of foods has been widely practised as a means of stabilizing raw materials and processed products that are stored under ambient conditions. Dried products and ingredients are a food category on their own (e.g. baked and pasta products, dried fruits, instant powders, spices). Since heating for water removal requires energy and alters quality, milder alternative methods have been developed. A range of products are stabilized at intermediate moisture contents by control of the water activity via partial water removal, addition of solutes (sugar or salt), adjustment of pH and/or addition of preservatives.

Technologies	Current status	Potential developments
Dehydration: Means of preserving food in a stable and safe condition by drastically reducing the water activity	Widely used technology spanning from sun-drying to controlled, cabinet-drying. Applied to a wide range of foods. Extensively used in LDCs to process local fruits, vegetables, fish and meat	Improvements in drying technologies for energy savings (Strumillo, 2006). High-quality drying, a major alternative for SMEs in LDCs to add value to raw materials. Growth in demand for high-quality dehydrated fruits and vegetables (Zhang *et al.*, 2006)
Spray drying: Transforming a liquid feed into a dry free-flowing powder using hot air	Still unchallenged drying technology for liquid foods (milk, extracts) resulting in many convenient products (Barbosa-Cánovas and Vega-Mercado, 1996)	Applications in the natural extracts and functional food industries are likely to increase. Alternative for encapsulation of valuable ingredients. Appropriate for SMEs in LDCs
Freeze-drying (FD): Removing water from the frozen state (ice) under vacuum	Recognized as the best drying technology for high-quality food products and preservation of cell viability (e.g. drying of probiotics). Process is slow, energy demanding and expensive	New niches for premium dried foods with better colour and flavour, particularly fruits and vegetables (e.g. use in breakfast cereals, gourmet soups). Developments in low vacuum (atmospheric) FD
Microwave drying: Focusing energy directly into the interior of the products	Already used in a number of non-food industries, provides shortened drying times, reduced costs and high product quality. Successfully adopted for drying of pasta products	Applications in food industries that require short drying times and higher throughput at the expense of higher capital and energy costs
Other drying methods	Novel forms of drying, such as the use of radio-frequency, super-heated steam and heat pump drying, are actively being investigated	Improved drying technologies (hybrid technologies) for energy savings and better quality foods with higher market value (Chua and Chou, 2005)

Technologies	Current status	Potential developments
Osmotic dehydration: Use of concentrated solutions to remove significant amounts of water	Extensively researched in the past 2 decades, yet few industrial applications (restricted to fruit pieces) (Raoult-Wack, 1994). Suggested as a pre-drying step for fruits (using sugar solutions) and vegetables	Attractive low-tech method but requires proper handling of waste solutions. Overall energy efficiency and environmental impact need evaluation. Impregnation of nutrients and flavours
Concentration: Partial removal of water from liquid foods by evaporation	Well-established technology as a cost-effective pre-drying step in dehydration of liquid foods. Utilized for bulk transport of frozen concentrated juices	Alternatives to evaporation may find increased applications in high-quality foods with improved aroma and flavour. Examples: freeze-concentration, pervaporation
Hurdle technologies: Combination of several preservation factors at low levels	Traditional preservation technologies that yield ambient-stable, intermediate moisture fruits, processed meats and dairy products (Leistner and Gould, 2002)	Possibilities for novel products using mild heating technologies and 'natural' preservatives (spices or herbs). Pre-processing alternative for bulk storage of fruits and vegetables in LDCs
Minimal processing (MP) technologies: Shelf-stable fresh foods with minor changes in freshness	Minimal processing technologies for fruits and vegetables may include a washing/disinfection step followed by addition of inhibitors of adverse reactions and modified atmosphere packaging (Ahvenainen, 2000; Artes and Allende, 2005)	Favoured by increased demand for fresh-like fruits and vegetables for healthy nutrition, convenience and quality. Alternative disinfectants (antioxidants, ozone, irradiation, organic acids, etc.) and packaging options come into scene (Rico *et al.*, 2007)

Heating/cooling/freezing. Heating technologies are increasingly aimed at delivering energy directly into the product. Fast removal of latent (freezing) and sensible heat from products after thermal processing on the spot (chilling) or after harvest or capture (e.g. aquatic foods) is receiving increased attention, may require use of sustainable energy alternatives. As is the case of other preservation methods, major impacts on food quality and safety are expected when several of these technologies become integrated simultaneously or sequentially into processing lines.

Technologies	Current status	Potential developments
Microwave (MW) heating: MWs penetrate within a food, heating the interior more rapidly	Applications already extensive in home appliances. Although moderately used industrially (e.g. in food tempering) MW volumetric heating often provides a higher quality product, increased production rates and energy savings that may offset higher capital costs	Combination of MW and conventional heating will become increasingly important for energy- and time-consuming processes

continued

Annex 1: Continued.

Ohmic (OH), infrared (IR) and radio-frequency heating	These and other heating technologies have been investigated for sterilization/pasteurization, drying, roasting, etc. Some are implemented on a commercial scale: IR heating in Japan (Sakai and Hanzawa, 1994) and OH for sterilization of particulate food products	Consolidation of emerging heating technologies is expected. Most of them provide superior quality foods in specific applications (e.g. pasteurization of viscous products). High potential in combination with other heating methods (e.g. drying)
Vacuum cooling (VC)	Vacuum (evaporative) cooling is presently used mostly for fast pre-cooling of horticultural products to promote extension of shelf life and improvement of product quality and safety	Integration of VC to processing of other foods (e.g. ready meals) to improve safety is attractive, but requires assessment of possible adverse effects on product quality (Zheng and Sun, 2004)
Chilling: Rapid lowering of the temperature of foods to less than 8°C	Applications range from raw materials (fruits/vegetables, fish, meat) to prepared foods (e.g. *sous-vide*). High-quality, convenient chilled prepared foods are growing at the expense of the frozen food category. Chilled products require delicate temperature control	New systems engineered for fast cooling of produce and fish (e.g. slurry ice systems) likely to find applications for export markets. Need improved logistics for temperature control throughout the distribution chain
Freezing: Drastic reduction of temperature below −18°C with conversion of water into ice	Frozen products are amply recognized as healthy, safe and convenient. Growth in some segments (seafood, IQF berries) is offset by decline in others (ready meals). Market opportunities arise in LDCs as incomes increase and distribution networks of frozen foods become available	Opportunities for SMEs in high growth segments of frozen fruits (e.g. berries) and vegetables. Faster freezing rates for higher quality using cryogenic fluids. Pressure-freezing appears attractive for high-quality (but expensive) frozen foods (Sanz and Otero, 2005)

Separation and Ingredients Technologies. Several processes are used to release valuable components from raw materials, separate them from the original matrix and produce concentrated/purified products in an appropriate form (concentrates or powders). Small- and medium-size industries in LDCs have a major opportunity to market extracts of bioactives from lower quality fresh raw materials and by-products for the functional foods market. Progress in separations science will lead to economically feasible processes that make available refined and functional food ingredients to replace or complement traditional raw materials.

Separation technologies

Technologies	Current status	Potential developments
Pre-extraction processes: Controlled disintegration of cellular tissue for efficient release of components	Grinding and pressing are conventional technologies in the sugar, oil and protein industries as well as in small operations	Prevalence for technologies proving a faster, more efficient and/or selective release using, for example, enzymes, ultrasound, HP, MW, extrusion, PEF (Wang and Weller, 2006)
Extraction processes: Liquid extraction of solutes from solid substrates using water or organic solvents	Conventional technology to face limitations on type of solvents that can be used to extract foods and contents of residual solvents	Use of aqueous solvents (e.g. water/ethanol) in oil extraction, applications of membrane technology to recover water or other solvents; more efficient use of solvents
Supercritical extraction (SCE): Extraction with compressed gases having high diffusivity and solvent capacity	Interest in SCE using carbon dioxide for safety and processing (fractionation). Presently used to decaffeinate coffee and tea, and in the extraction of hops. High capital costs involved	Superior quality flavours, spices and essential oils by SCE have been extensively studied and some products are in the market (Brunner, 2005). Potential exists for the isolation of natural preservatives (Reglero *et al.*, 2005). Other applications in detoxification
Membrane separation (MS): Use of porous membranes to fractionate food components	Membrane technology (micro-, ultra- and nanofiltration) already established in many food industries. Ample availability of membranes (e.g. pore sizes) and absence of heating provide unique opportunities for fractionation	Larger demand for more convenient and purified functional ingredients (e.g. fractionated whey proteins) (Kulozik, 2008). More stringent regulation for cleaner effluents from processing plants
Chromatography: Separation at molecular level by affinity or size	A powerful separation technique already used at large scale to separate glucose from fructose	Novel and cheaper sorbent materials could greatly advance applications of chromatographic separations in the food industry

continued

Annex 1: Continued.

Ingredients technologies

Technologies	Current status	Potential developments
Product technologies: Conventional technologies leading to food extracts	Extracts are convenient, standardized and concentrated forms of ingredients. Conventional technologies for production of concentrates and dry powders well known around the world. Some emerging countries have a long tradition of exporting plant and algal extracts (e.g. India)	Conventional extracts may become valorized when transformed into high-quality ingredients. Functional plant extracts and nutraceuticals are in high demand by the prepared food and cosmetic industries. Market niches offer added opportunities
Microencapsulation: Components trapped within an edible matrix (wall) and delivered as small particles	Conventional encapsulation (e.g. liquid flavours) already familiar in the food industry. New microencapsulation technologies provide chemical protection and functionality to valuable compounds (e.g. a liquid flavour) and to beneficial bacterial cells (e.g. probiotics) (Holmgren, 2006)	Microencapsulation technology advances at fast pace in designing encapsulating matrices (solid or liquid) for specific types of fat-soluble, water-soluble compounds and bioactives. Use in tailoring nutrient delivery systems (Augustin and Sanguansri, 2008)

Transformation (Conversion) and Structuring Technologies. The bulk of the processed foods industry involves transformation and structuring of food components during the production process. Dairy products, baked goods, processed meats, snacks, sauces and dressings, among others, are structured products. This is the most important category of processed foods in terms of the size of the industry (Bruin and Jongen, 2003).

Technologies	Current status	Potential developments
Emulsification: Dispersing immiscible phases as fine droplets within a continuous phase	Many structured foods are emulsions formed by mechanical shearing devices such as high-speed agitators and homogenizers (e.g. mayonnaise, salad dressings, etc.)	Membrane emulsification may find increased application to form stable emulsions with lower energy consumption and less use of surfactants
Extrusion: Use of a screw extruder to continuously cook, pasteurize and shape starchy flours and protein meals into dried products	Established technology around the world. Advantages are versatility in use, energy efficiency, high productivity, low operational costs and absence of wastes and pollutants	Extrusion is likely to increase its importance in LDCs to produce pre-cooked flours, infant foods, fish feeds, etc. Applications in SMEs (e.g. precooking, decontamination, etc.) may only require low-cost versions of extruders

Technology	Status	Comments
Gelation: Trapping abundant water in semi-solid foods using hydrocolloids	Gels already important products in the dairy, meat and fish industries, and in gastronomy. The source of gelling materials is expanding	Gelation may become important in the design of new food structures, dietetic foods (e.g. reduced calorie foods, control of satiety) and delivery systems
Frying: Immersing food pieces in hot oil (deep-fat frying) to impart unique textures and flavours to foods	Used extensively worldwide in snacks production, food outlets (including street vendors in LDCs) and at home. Fried products are high-calorie foods due to oil absorption during frying	Concern about calorie density, type of fats, the generation of toxic acrylamides and polymeric substances, and interest in preserving the nutrient content, flavour and colour of raw materials are likely to change frying operations (e.g. vacuum frying)
Enzymic processing: Promoting the action of natural or added enzymes for process improvement and product functionality	Well-understood and extensively used technology in the dairy, juice, sweeteners, starch and flavour industries, where enzymic processing has proven to be cost-effective	Natural and GMO-derived commercial enzymes for biotransformation of fats, carbohydrates and proteins. Improved enzymic reactors for biotransformations, cross-linking enzymes for food structure build-up
Micro-engineering: Use of devices developed for microtechnologies	Microdevices or arrangements of capillaries or channels <1 mm that deal with small amounts of fluids (10^{-6}–10^{-9} l) are currently only available at laboratory scale	Microfluidic devices have potential to significantly change the fabrication of dispersed food systems (emulsions and foams) (Skurtys and Aguilera, 2008). Possible applications in the design of analytical devices
Nanotechnologies: Emerging technologies to manipulate materials, devices and products at dimensions in the order 10–100 nm	Great hopes but so far few products and applications. Already consumer groups worry about safety and environmental impact of nanotech products to be used in foods	If present concerns are overcome food nanotechnology is likely to have a major impact in areas such as smart sensors in packaging, self-cleaning surfaces, encapsulated delivery systems, nanoemulsions and food nanoparticles (Sanguansri and Augustin, 2006; Lee *et al.*, 2008)

continued

Annex 1: Continued.

Packaging technologies. Packaging technologies will aim at providing added protection and information to consumers while reducing the amount of packaging used, promote recycling and facilitate disposal. Increased use of novel, environmentally friendly materials derived from natural sources (e.g. bio-based packaging) is expected. Package safety from migrating substances will continue to be of concern. Most packagers in LDCs are SMEs looking for new technologies offering reliability and lower processing costs.

Technologies	Current status	Potential developments
Active packaging: Package, product and the environment interact to extend shelf life	Several active components already in use (sachets or 'in the wall') to improve safety and sensory quality (ethylene and O_2 scavengers, antioxidant and antimicrobial emitters, etc.)	Introduction of novel active components (e.g. nanoparticles, enzymes, etc.), but should overcome higher cost, legislative restrictions, consumer safety and environmental impacts (Vermeiren *et al.*, 1999; Lopez-Rubio *et al.*, 2006)
Intelligent packaging: Packaging detects, senses, records and provides information about the contents (Yam *et al.*, 2005)	Truly 'intelligent' packaging still at the concept level. Often confused with smart packaging (see below)	Among others, it is expected that packaging could detect and signal the presence of microbiological growth in packaged foods (safety) or aromas (quality)
Smart packaging: Packages with attached devices that can store and transmit data	Already in use for logistics (barcodes, radio-frequency identification devices – RFID) and consumer information (time–temperature indicators – TTI)	Need of less-expensive devices for wide use in traceability, improved logistics and added convenience (e.g. appliances interacting with encoded information for food preparation)
Edible barriers and films: Starch, protein and fat-based barriers and films that protect products	Emerging alternatives to stabilize some fresh and processed foods to satisfy consumers' demands for longer shelf-life products (Bourlieu *et al.*, 2008)	Wide-ranging applications (e.g. to prevent moisture migration and microbial contamination) in many food categories. Type and availability of effective films will expand
Modified atmosphere packaging (MAP): Altering the gases surrounding a product or commodity to extend the storage life	MAP extends shelf life and preserves the high quality of foodstuffs. Concerns about safety due to growth of pathogenic bacteria and in quality due to fermentation	Expected to increase in use since it benefits consumers (e.g. more stable fresh products) and provides greater flexibility in production and within the distribution chain

Technologies	Current status	Potential developments
New packaging materials: Mostly related to biodegradable and bio-based materials (BBM) and composites	Few commercial BBM for food packaging (starch- and microbial polymer-based) presently available. PLA and PHB hampered by cost and low performance	Future of BBM associated with performance (barrier), processing and cost. Provide much lesser environmental impact than plastics. Nano-biocomposites develop in an uncertain scenario (Sorrentino et al., 2007)

Storage and distribution technologies. Postharvest losses in developing countries may be reduced by controlling the temperature and/or moisture of grains, horticultural produce and fish. Transfer of simple technologies, adoption of better practices at farm level and improved marketing channels are crucial. A major contribution to food authenticity and traceability is expected to come from in-line inspection using non-destructive, non-invasive spectroscopic and imaging techniques. Efficiency and traceability in distribution systems stimulate use of robotics and information technologies (ITs).

Technologies	Current status	Potential developments
Storage systems	Conditions for minimizing physical and quality losses of food raw materials after harvests are well established	Improved packaging materials and containers. Underground sealed storage of grains. Storage under gas atmospheres (nitrogen, ozone, etc.)
Cold storage and transport: Transport and storage at reduced temperature (frozen and chilled foods)	An established technology in the developed world that uses a network of storage facilities served by millions of refrigerated road vehicles (James and James, 2006)	Developing countries implement the infrastructure for the refrigerated/frozen food chain aimed at export markets. Advances in monitoring in-transit and stored products by IT
Robotics: Use of robots in the production and distribution processes	Robots used in specific production systems (e.g. confectionery industry), and in areas of materials handling and secondary/tertiary packaging operations (Wallin, 1997). Challenge in delicate nature of many food materials	Holds out a promise of reducing costs by increased consistency, more efficient production and reducing labour costs. May also find applications in hostile production environments
In-, on-, at-line quality control: Application of non-destructive non-invasive technologies for quality control	Spectroscopic techniques and electromagnetic probing demonstrating opportunities for fast on-line analysis of food components and detection of surface and internal defects	Advances need to be made to bring down the costs of these devices so that they become affordable for continuous, on-line quality monitoring by the food industry (Scotter, 1997; Sun, 2004; Cen and He, 2007)

Annex 2: Summary of Bioinformatic Databases

Database name	Type of information	Uses of information	Database host	Web site URL [a]
NCBI GenBank & PubMed	DNA sequences (GenBank); protein sequences; genome structure; publications	Species, gene & protein ID; genomic comparison	National Centre for Bioinformatic Information	www.ncbi.nlm.nih.gov/
DDBJ (DNA DataBank of Japan)	DNA (as GenBank)	Species & gene ID; genomic comparison	National Institute of Genetics	www.ddbj.nig.ac.jp/
EMBL (European Molecular Biology Laboratory)	DNA (as GenBank), protein sequences; microarrays; bioinformatic tools	Species ID; microarray analysis; protein function & structure	European Bioinformatics Institute	www.ebi.ac.uk/embl/
FishTrace	Cytochrome b & rhodopsine gene sequences from 200+ European fish species	Fish species ID	JRC – Joint Research Centre, Italy (EU consortium)	www.fishtrace.org
FishBase	General information on 30,000 world fish species	Fish classification, uses, habitat, identification, images, genetics on fishes	Consultative Group on International Agricultural Research (CGIAR)	www.fishbase.org/search.php
Gramene	Rice, maize & grasses gene, protein & metabolite info	Marker (QTL) ID; traits; genetic diversity of grasses	US collaboration	www.gramene.org/
GrainGenes	Wheat, barley (cereals) gene, protein & metabolite info	Marker (QTL) ID; traits; genetic diversity of cereals	USDA (US Department of Agriculture.) (US consortium)	wheat.pw.usda.gov/GG2/index.shtml
OlivTrack	Olive genetic markers	Olive oil traceability	University of Parma (EU consortium)	www.dsa.unipr.it/foodhealth/oliv-track/index.html
PLEXdB database	Plant & plant pathogen gene expression data	Plant functional genomics; microarray data from maize, barley, grape, rice & tomato	Iowa State University (US consortium)	www.plexdb.org/index.php
ArkDB	Genetic information on 12+ farmed species	QTL mapping of animal, bird & fish species for breeding	Roslin Institute	www.thearkdb.org/

Database	Description	Application	Institution	URL
Livestock Genome Mapping Programmes	Genetic information on six livestock species	QTL mapping of animal, bird & fish species for breeding.	INRA, France	locus.jouy.inra.fr/cgi-bin/bovmap/livestock.pl
NAGRP (National Animal Genome Research Program)	Genomic information from various livestock species	QTL mapping of animal, bird, fish & crustacean species for breeding	USDA (US consortium)	www.csrees.usda.gov/nea/animals/in_focus/an_breeding_if_nagrp.html
Animal genomics community web site	Compendium of databases from 14 animal species	Database searches for animal genetics information	Iowa State University	www.animalgenome.org/community/other.html
Organism-Specific Genome Databases	Compendium of eukaryote, prokaryote & viral databases	Database searches for animal, plant, fungal, microbial & viral info	University of California	restools.sdsc.edu/biotools/biotools10.html
Sanger Institute	Genome sequences	Genome mapping & sequence comparison	Sanger Institute	www.sanger.ac.uk/Projects/
The J. Craig Venter Institute (JCVI)	Compendium of microbial, fungal & plant genetic info	Functional genomics; QTL & genome mapping; sequence comparison	J. Craig Venter Institute	www.tigr.org/ www.jcvi.org
MICADO (Microbial Advanced Database Organization)	Microbial genomes	Gene mapping & functional analysis	INRA, France	genome.jouy.inra.fr/cgi-bin/micado/index.cgi
Barcode of Life	DNA sequences of the cytochrome oxidase subunit 1 (COI) gene	Taxonomic classification; sample identification	National Museum of Natural History, USA	www.barcoding.si.edu/ www.barcodinglife.org/views/login.php?&
HapMap project	Database of human genes	Gene expression; genetic variation & markers	Cold Spring Harbor Laboratories, USA (international consortium)	www.hapmap.org

[a]All databases accessed on 13-05-09.

5 Enabling Environments for Competitive Agro-industries

Ralph Christy,[1] Edward Mabaya,[2] Norbert Wilson,[3] Emelly Mutambatsere[4] and Nomathemba Mhlanga[5]

[1]Professor, Department of Applied Economics and Management, Cornell University, Ithaca, New York, USA; [2]Researcher Associate, Department of Applied Economics and Management, Cornell University, Ithaca, New York, USA; [3]Associate Professor, Department of Agricultural Economics, Auburn University, Auburn, Alabama, USA; [4]Evaluation Analyst, African Development Bank, Tunis, Tunisia; [5]PhD Candidate, Department of Applied Economics and Management, Cornell University, Ithaca, New York, USA

Introduction

In the most recent decades, developing nations have focused predominantly on economic prescriptions for 'getting markets right' by adjusting macroeconomic policy, privatizing state-owned enterprises and opening domestic markets to international trade in agricultural commodities and currencies. The implicit assumptions are that 'structural adjustment programmes' will attract foreign capital through the domestic and international private sectors, thereby making domestic industries more competitive. As a result, a large part of economic development policy has centred on creating 'enabling environments' for which capital would then be attracted to invest in both general market-based solutions and specific firm strategies that contribute to the economic growth and development goals of the nation. The effectiveness of those policies has varied as the interrelationships among macroeconomic policies, firm strategy and social goals all hinge on different analytical frameworks.

A defining characteristic of most developing economies is the relative importance of agriculture in their national economies and consequently the design of alternative strategies employed to achieve national goals. The green revolution fuelled rapid growth of agricultural productivity in Asia and this, coupled with investments in infrastructure, reduced rural poverty dramatically. Advances in economic development in Latin America, however, occurred in a tiered policy structure that favoured promoting value added, agriculture-based industries and niche products for export markets with convertible foreign currencies. While Asia and Latin America identified strategies to stimulate economic development, sub-Saharan Africa (SSA) placed greater emphasis on the political economy at the expense of economic growth and development, thereby leaving fewer resources to overcome key rural economic development problems such as persistent poverty, shortage of preventive health care, fragmented infrastructure and food insecurity. The gap between the national economies in SSA countries and those of developed countries has widened, while the gap has narrowed for the emerging and competing regions of Latin America and South-east Asia. What has become abundantly clear in this post-structural adjustment era is that institutions matter. The reliance on markets in achieving policy goals requires developing nations to invest in institutions and services that will allow markets to function well in a global economy.

Translating market-based policies in a global economy into desired social and economic ends requires a fundamental shift in thinking about a cornerstone concept, *comparative advantage*, which explains international trade, to an expanded understanding of global markets that is based on competitiveness. Today nations are advancing policies to enable firms and industries to gain a competitive advantage in global markets, as opposed to the conventional (Ricardian and Heckscher-Olin-Samuelson) international trade frameworks. In the literature, the concept of competitive advantage has a number of distinct meanings. Herein, we use the concept in a manner similar to that defined by the Organization for Economic Cooperation and Development (OECD) as 'the degree to which, under open market conditions, a country can produce goods and services that meet the test of foreign competition, while simultaneously maintaining and expanding domestic real income' (OECD, 1992, p. 237). An important aspect of the concept of competitive advantage lies in the efforts of many organizations to provide measures of an enabling economic environment to foster competitiveness.

For policy makers identification of those elements is complicated by an underlying feature of the globalization process – the rapidity of change occurring within and across national economies. The globalization process has the potential to benefit emerging economies, as proven by the remarkable rates of economic growth in many parts of the world. This process has fused the theoretical stages of economic development and raised the premium on the traditional, sequential approach, which holds that the state must first create an enabling environment, which is then followed by private-sector investments. Emerging market governments are investing in the necessary components to foster economic stability, including infrastructure, open telecommunication markets and Internet-based distance learning programmes while, at the same time, competition for capital, driven by the rapidity of the globalization process, is causing multinational corporations to aggressively seek opportunities to capture higher returns. As a result, the private sector in many cases is not necessarily waiting for an enabling environment to be created by the state. Rather, it is working in partnership with governments to develop a suitable environment, while simultaneously pursuing market opportunities.

Efforts by policy makers to measure competitiveness, enhance enabling environments and comprehend the rapidity of change in domestic and global economies have given rise to three fundamental questions, which are addressed sequentially in this chapter:

1. How well do measures of business climate by international organizations and research institutions relate to the competitiveness of mostly agricultural economies?

2. If current measures are inadequate, what are the essential factors underlying agro-industry competitiveness in developing countries?

3. How can we reform the public process, in the context of radical change, to develop and enact creative policies to improve the relative competitiveness of agro-industries in emerging markets?

Business Climate Assessments

How well do measures of business climate by international organizations and research institutions relate to the competitiveness of agrarian economies?

The assessment of business climate (or environment) dates back to the late 1970s, when the World Economic Forum started publishing the *Global Competitiveness Report*, with the assessment and ranking of economic competitiveness for 16 European and North American countries. The level of interest in these climate assessments grew considerably from then on, particularly in the past decade, as increased globalization created demand for methods of providing signals to investors interested in foreign direct investment (FDI). In addition to this main use, business climate assessments have also been seen as a way of inspiring reform, a direct result of the ranking system used in these analyses. In most cases, focus is placed on those economies whose rankings have improved substantially from one year to another through special mention in case studies or reform awards. The assessment of business climates is usually accompanied by recommendations on which procedures to reform, the adequacy of which for agribusiness and agro-industries will be discussed later in this chapter. A general overview and appraisal of business climate measures is presented next.

The interest in business climate assessments has resulted in a proliferation of measures to quantify the suitability of given economic environments for sustainable business development. Figure 1 presents the chronological order of the

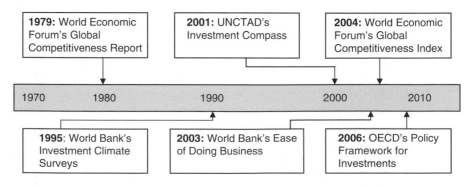

Figure 1. Timeline of competitiveness and investment climate assessment frameworks.

various *competitiveness* and *investment climate* indices that have been developed over the last 3 decades. Major international organizations now produce and publish one or more indices for use in measuring the level of competitiveness. The most widely monitored and applied indices are the Ease of Doing Business Index (World Bank, 2003) and the Global Competitiveness Index (GCI) (World Economic Forum, 2004). Several others also exist, including the World Economic Forum's Growth Competitiveness, Current Competitiveness and Business Competitiveness indices; the United Nations Conference on Trade and Development's (UNCTAD) Investment Compass (IC), Global Investments Prospects Assessment and Inward FDI Performance Index; the Policy Framework for Investments (OECD, 2006); and Investment Climate Surveys (World Bank, 1995). Instruments such as governance indices also are thought to be good indicators of investment climate, and they, too, are used to study how institutional environments impact on business performance. The next subsections discuss these measures in greater detail and provide a critique of the measures with specific attention paid to the appropriateness of their use in the agribusiness industry (summarized in Table 1).

Ease of Doing Business Index

The Ease of Doing Business Index is a relative measure of how well the business climate facilitates efficiency in ten stages of a business's life, i.e. starting a business, dealing with licences, employing workers, registering property, getting credit, protecting investors, paying taxes, trading across borders, enforcing contracts and closing a business. The index is calculated from a three-stage process that involves the following:

- ranking of the various components of each stage according to a predetermined scoring framework as outlined in Appendix 1;
- computation of the stage-level ranking, defined as the average of the percentile rankings of the stage's component indicators; and
- computation of the overall business climate ranking for the country – the average of the percentile rankings of each of the ten stages.

The final country score, which describes its position in the percentile ranking relative to other world economies, will take a value between 1 and the total number of countries included in the assessment. Appendix 2 shows an example of how the Ease of Doing Business Index was computed for Egypt, the top economic reformer in 2006/07 according to World Bank's (2007) *Doing Business 2008* report.

One of the main merits of the Ease of Doing Business Index is that it covers, in depth, most regulations directly affecting business operation. It is a good attempt to determine what regulation constitutes binding constraints, which reform packages are most effective and how these issues are shaped by the country context. A few limitations can also be identified. First, the assessment is limited in scope to cover only those regulations directly linked to business operations, but it says little about the effects of other (often equally important)

Table 1. Competitiveness assessment frameworks and their limitations.

Index and year established (source and coverage)	Key strengths	Limitations in assessing agro-industry competitiveness
1979: World Economic Forum's Global Competitiveness Report 125–131 countries	Holistic – considers policies, factor endowments, and institutions Adjusted for each country's level of economic development	Does not inform policy at industry or sectoral level Fails to capture value chain components
1988: UNCTAD's Inward FDI Performance Index 141 countries	Uses 3-year periods to offset annual fluctuations in the data Captures influence of all factors other than market size	Does not inform policy at industry or sectoral level Neglects domestic investment, hence key components of agro-industries
1995: World Bank's Investment Climate Surveys 50 countries	Adapted to country context and sector priorities Covers business perceptions	Small sample sizes limit global comparisons Places emphasis on foreign investment
2001: UNCTAD's Investment Compass	Focuses on both availability and quality of infrastructure Land tenure central to the analysis Considers objectives of investors and policy makers	Does not inform policy at industry or sectoral level Fails to capture value chain components
2003: World Bank's Ease of Doing Business Index 178 countries	Highly comprehensive – covers major business considerations from inception to folding of operations Provides an excellent assessment of regulatory framework Allows inter-country comparisons	Disregards broader business environment, e.g. macroeconomic fundamentals Biased in favour of formal establishments Is not sector-specific
2004: World Economic Forum's Business Competitive Index 121 countries	Examines firm-level efficiency	Intensive data requirements
2004: World Economic Forum's Global Competitiveness Index 131 countries	Holistic – considers policies, factor endowments, and institutions Adjusted for each country's level of economic development Provides a policy evaluation criterion	Does not inform policy at industry or sectoral level Fails to capture value chain components
2006: OECD's Policy Frameworks for Investment	Directly targets policy makers Highlights key areas for private-sector-led growth-enabling proactive policies	Focused on foreign investment Is not sector-specific

policies and institutional arrangements, such as the effect of macroeconomic policies, quality of infrastructure, proximity to markets, security of property, gender biases in business regulations and transparency of government procurement. In its country ranking the index is biased in favour of formal industries, as higher rankings tend to be associated with economies that experience more formal sector growth, more jobs and a shrinking share of the economy in the informal sector. Moreover, although conclusions drawn from the analysis are taken as non-sector-specific, to make the data comparable across countries, the Ease of Doing Business Index refers to specific types of business – drawing data from a limited liability company operating in the largest business city. Sectors lying outside this class are excluded, even though these may constitute larger proportions of business activity in specific countries. No direct assessment is made of the quality of downstream or upstream market environments, and their subsequent impact on the efficiency of business operations at manufacturing level (the index's target).

Global Competitiveness Index

Global competitiveness assessments have been performed by the World Economic Forum since 1979 and have evolved over the years in tune with the changing international environment. The most recent development, the GCI, resulted from the work of Xavier Sala-i-Martin in 2004 (World Economic Forum, 2004). In addition to the GCI, the World Economic Forum has produced three other measures of competitiveness: the Growth Competitiveness, Current Competitiveness and Business Competitiveness indices, discussed briefly below. The GCI provides an overview of critical drivers of productivity and competitiveness, categorized into nine pillars: institutions, infrastructure, macroeconomy, health and primary education, higher education and training, market efficiency, technological readiness, business sophistication and innovation. Data from the Executive Opinion Survey, combined with hard data, are used to rank countries on each component of the index. The pillars can be integrated into three broad categories, namely basic requirements, efficiency enhancers, and innovation and sophistication, as specified in Appendix 3.

The GCI is computed for a sample of 125 countries of varying degrees of economic progress. To highlight the differences in policy priorities for economies at different levels of economic progress, the index is designed so that it assigns higher weight to those pillars key to the specific phase/position of a particular national economy. For factor-driven economies (countries that compete based on factor endowments), for example, more weight is given to basic requirements relative to efficiency enhancers, and much less to innovation. Innovation-driven economies, on the other hand, compete with new and unique products and therefore rely more on innovation and sophistication factors for competitiveness. The division of countries into the three broad categories (or stages of development) is explained in Box 1.

One of the major advantages of the GCI is its holistic nature, stemming from understanding national competitiveness as a set of factors, policies and

Box 1. Weighting according to stage of development. (From World Economic Forum, 2006.)

	GDP per capita (US$)	Basic requirements (%)	Efficiency enhancers (%)	Innovation factors (%)
Factor driven	<2,000	50	40	10
Transition from stage 1 to 2	2,000–3,000			
Efficiency driven	3,000–9,000	40	50	10
Transition from stage 2 to stage 3	9,000–17,000			
Innovation driven	>17,000	30	40	30

institutions that determine the level of productivity in a country. Productivity implies making better use of scarce resources; therefore, competitive economies are expected to experience higher growth rates (World Economic Forum, 2006). Unlike the Ease of Doing Business Index, which focuses only on those factors directly affecting the business environment, the GCI also recognizes the importance of the macroeconomic environment, human development, market efficiency, technology and innovation in developing and sustaining global competitiveness. These differences give rise to different competitiveness ranks. Note that, while Singapore ranks first in 2006 according to the Ease of Doing Business Index, lower rankings in health, education and business sophistication push it to the fifth position in the GCI ranking, where Switzerland ranks first. The GCI's component rankings also are modified according to each country's level of economic development, thus avoiding the 'one-size-fits-all' pitfall common to most competitiveness ranking criteria.

The main limitations of the index are, first, that it focuses on macroeconomic features and offers no industry or sector-specific appraisal of the operating environments. Second, the discussion on market efficiency fails to capture the critical value chain components that enhance or hinder competitiveness in any given industry. Third, the issue of geographic location of firms and proximity to markets also remains unaddressed.

The Growth Competitiveness Index

The Growth Competitiveness Index is a characterization of a set of institutions and economic policies that support high rates of economic growth over the medium term, i.e. over the coming 5 years. The index is based upon a country's performance in three areas (sub-indices): level of technology, quality of public institutions and macroeconomic conditions related to growth. Dividing world economies according to innovation into core and non-core economies (patenting is used as a measure of innovation capacity), the index assigns different weights to factors within the sub-indices for economies with different innovation capacities. For example, for core economies, weights of ½, ¼ and ¼ are assigned to technology, quality of public institutions and macroeconomic conditions, respectively, whereas weights of $1/3$ each are assigned to the three sub-indices for non-core economies. The GCI, which grew directly from the Growth Competitiveness Index, attempts to incorporate many factors that drive productivity into a broader measure of competitiveness. The World Economic Forum currently publishes both indices.

Current Competitiveness Index

The Current Competitiveness Index (CCI) evaluates the underlying conditions that define current level of productivity (World Economic Forum, 2000). It uses macroeconomic indicators to estimate the set of institutions, market structures and economic policies supportive of high current levels of prosperity, thus giving an indication of the effectiveness with which an economy utilizes its current stock of resources. The CCI is based upon two sub-indices that focus on firm sophistication and qualities associated with the national business environment, drawing on a complex array of variables with a demonstrated statistical relationship to gross domestic product (GDP) per capita. The additional value of this index is that it goes beyond examination of aggregate variables to capture the microeconomic conditions that support a high level of sustainable productivity. The CCI was a precursor to the Business Competitiveness Index (BCI), discussed below.

The Business Competitiveness Index

The BCI developed by Michael E. Porter ranks countries by their microeconomic competitiveness. The focus on the microeconomy emerges from the realization that, although sound macro policies create potential for improving national prosperity, wealth is created by firms, and it is the efficiency with which firms operate that drives firm, industry and national competitiveness. The BCI thus highlights, in detail, the microeconomic underpinnings of competitiveness, emphasizing a range of company-specific factors conducive to improving efficiency and productivity (World Economic Forum, 2007). Competitive strengths and weaknesses are identified in terms of a country's business environment conditions and a firm's operations and strategies. These features are then used to assess the sustainability path of the country's current prosperity.

Productivity is thought to be driven at the micro level by the sophistication with which locally based firms compete and by the quality of the micro-business environment in which they operate. The BCI is based upon two sub-indices that focus on these two foundations. Factors that drive company strategy include production processes, staff training, marketing, capacity of innovation, branding and value chain, whereas those that affect business environment include factor conditions, demand conditions, nature of related and support

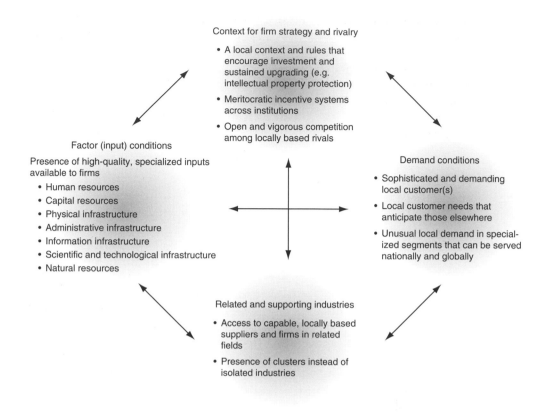

Figure 2. The microeconomic business environment. (From Porter, 2004.)

industries, firm strategy and domestic rivalry (*see* Appendix 4). The factors that affect microeconomic business environment can be summarized into the 'national diamond' as characterized in Figure 2. Because the impact of individual factors or variables is difficult to isolate, the common factor analysis is used to compute the sub-indices, which are then averaged to obtain the overall BCI value. Variable data are obtained primarily from the Executive Opinion Survey; quantitative measures are used where data are available, e.g. for patenting rates and Internet and cellular telephone penetration. Due to data intensity, the BCI is currently published for only 121 countries.

As a measure of competitiveness, the BCI has the clear advantage of focusing on the *local* business environment, the environment of primary relevance to firms. The Index is based upon the realization that, whereas broad macroeconomic circumstances are able to provide the opportunity to create wealth, they do not themselves create wealth, making them only necessary but not sufficient conditions for competitiveness (Porter, 2004). The BCI computation framework enables also rigorous assessment of firm-level and country competitive strengths and weaknesses, thereby unmasking underlying factors affecting economic growth. Additionally, the framework recognizes differences in challenges and opportunities faced by economies at different levels of development, accounting for these in the assessment of competitiveness through the focus on the local environment.

The main limitation of the BIC is that it is ultimately an aggregated national index. Although the underlying data can be broken down to individual firms, industries and sectors (the index is computed using data drawn from over 7000 business surveys), no industry or sector-specific conclusions are drawn. Competitiveness is benchmarked to GDP per capita: the adopted measure of 'national prosperity', with the objective of understanding, for each nation, those microconditions significantly correlated to higher levels of prosperity. Also, the analysis is data intensive; imperfect data are often employed presenting an analytical challenge to the econometric modelling process.

Investment Compass

The IC, an innovation of UNCTAD, is a benchmarking tool specifically designed for developing countries for use in analysing the main economic and policy determinants that affect the investment environment. The database at present covers 55 developing countries. The IC comprises six key factors thought to influence the investment environment: resource assets, infrastructure, operating costs, economic performance and governance, taxation and incentives, and regulatory framework. The construction of the compass follows a multistage quantification process in which the key indicators are broken down into groups of variables, also broken down to measurable indicators (see Appendix 5). Each indicator is then assessed according to a rating system that takes values between 1 and 100. To obtain indicator ratings, a normalization process is used that involves fixing the minimum and maximum possible nominal values for each indicator and, through scaling, converts the nominal values to the 1–100 scale. The simple arithmetic average is used to aggregate the relevant normalized indicators into variables, and the variables into the key areas. Data on indicators are obtained from UNCTAD-administered special national surveys, foreign investment questionnaires and international statistics databases.

Like the Ease of Doing Business Index and the GCI, the IC provides an inter-country comparison according to performance in all investment areas. Additionally, the compass makes horizontal comparisons for each key area between countries, as well as vertical comparisons of performance in each key area for a given country. The key areas can be categorized according to ease of reform through policy action, from least responsive to policy action (resource assets) to most responsive (regulatory framework). Classification can also be made according to distinct objectives as viewed by policy makers, namely, production and employment, export development and technological 'catch-up' objectives, or according to investor objectives as viewed by foreign investors, i.e. domestic market-seeking, resource- or asset-seeking and export-oriented objectives.

Compared with other measures of business climate attractiveness, the IC highlights some indicators of greater significance to developing and transition economies. First, it addresses the effect of resource assets, particularly the effect of availability of raw materials such as minerals, agricultural commodities and energy reserves. The presence of a rich resource base can be important for explaining investment in areas where other determinants of business climate attractiveness are weak. Second, it addresses land ownership and transfer, specifically regulations pertaining to ownership of customary land and the nature of land titles. This approach is thought to have an impact on influencing market entry. Third, the IC considers the effect that the presence of preferential trade agreements with major markets has on investment decisions, an issue of particular importance when considering the flow of foreign investment to developing economies, and of importance to the agriculture sector. In terms of infrastructure, the compass highlights quality and access to the basic forms of infrastructure: access to water, mobile phones and road networks, which tend to be of greater relevance to developing countries.

Inward FDI Performance Index

The Inward FDI Performance Index is one of several indices produced by UNCTAD to measure how well economies perform in attracting foreign investments. It is calculated as the ratio of a country's share in global foreign direct investment (FDI) inflows to its share in global GDP; an index value greater than 1 implies that the country receives more FDI than its relative economic size. The index thus captures the influence on FDI of all factors other than market size, e.g. business climate, economic and political stability and presence of natural resources. The index is computed as a 3-year average to smooth out annual fluctuation in investment flows, dates back to 1988, and currently covers 141 countries. UNCTAD has published the Inward FDI Performance Index in the *World Investment Report* since 2001, together with other indices including the Inward FDI Potential Index and Outward FDI Performance Index. The Report, benefiting from investment data obtained from such sources as the World Investment Prospects Survey (1995 to date), provides a more detailed account of investment patterns as well as the driving forces behind observed investment trends.

UNCTAD also produces the Global Investment Prospectus Assessment (GIPA), a measure of the short- and medium-term prospects for foreign investment at the global, regional and industry levels. GIPA serves the same objective as the Policy Framework for Investment (described below), that is, to equip governments and, in this case, businesses with an instrument for proactive development of policies and strategies to influence future flows of investment. Employing data from three global surveys of transnational corporations, FDI experts and investment analysts, the assessment also evaluates evolving trends in strategies of transnational corporations, as well as FDI policies.

The FDI indices have the advantage that they employ easily accessible data and uncomplicated computation methods. The main limitation lies in the foreign investment focus, as opposed to an assessment of factors relevant to industry development in general. As indirect measures of business climate attractiveness, the FDI performance indices per se can only tell us where FDI tends to flow, without fully explaining the determinants or causal factors. Also, these are aggregated national indices that are neither industry- nor sector-specific.

Policy Framework for Investment

The Policy Framework for Investment was developed as an instrument for guiding policy reforms in critical areas of a country's economic environment – to mobilize private investments that support economic growth (OECD, 2006). The framework proposes guidelines on policy issues to be considered for reform by governments interested in creating an attractive environment for private-sector investments. Following the UN Monterrey Consensus on Financing for Development, the framework identifies ten policy areas as having the strongest impact on the investment environment: investment policy, investment promotion and facilitation, trade policy, competition policy, tax policy, corporate governance, policies for promoting responsible business conduct, human resource development, infrastructure and financial sector development and public governance. The guiding principles or logic and the specific issues to be addressed in each policy arena are summarized in Appendix 6.

The application of the framework is guided by three principles: (i) policy coherence; (ii) a transparent approach to policy formulation and implementation; and (iii) regular evaluation of the impact of existing and proposed policies on the investment environment. As a result, policy questions are designed to ensure an integrated approach to policy and investment environment interaction, to reduce uncertainty and risk for investors and to evaluate how well government policies uphold established good business practices.

The Policy Framework for Investment occupies a special niche in the business climate evaluation literature. Unlike the Ease of Doing Business Index or GCI, whose target audience is investors, the Policy Framework for Investment directly targets policy makers and attempts to equip them to focus on areas likely to produce the highest gains in private-sector growth. The framework, at present, does not attempt to rank economies according to current performance in addressing the outlined policy issues, but rather provides criteria through which governments can evaluate and improve the performance of their policies. In this strength also lies the main limitation of the Framework that it is simply a strategic approach for policy development – not an assessment tool. Thus, the Framework per se cannot be used to evaluate how conducive a country's policy environment is in facilitating or promoting investment relative to other competing investment destinations.

An important feature of the Policy Framework, observed to some extent in the first two policy fields, is that it focuses on those policy reforms necessary to attract foreign investments rather than those necessary to develop the private sector in general. Although these are usually consistent, the focus of the Framework speaks also to the ultimate goal of undergoing reform: to attract foreign investment.

Investment Climate Surveys

The Investment Climate Surveys are a production of the World Bank's Investment Climate Unit and a continuation of the World Bank's work from the mid-1990s to generate statistical information for formal investment climate assessments. The surveys use companies as the sampling unit and employ a written questionnaire administered through face-to-face interviews with company heads. Although surveys are generally adapted to the country context in sectors of coverage or thematic scope, depending on local priorities in policy reform and policy research, a core structure is maintained for consistency and comparability with international benchmarks. A standard sampling procedure is employed that involves using the business establishment (rather than the firm per se) as the unit of analysis, thus ensuring that each country covers a minimum set of common sectors and that within each country the major growth industries are adequately represented.

The written questionnaire comprises 12–15 sections of standard questions, 11 of which are core (see Appendix 7). The questions are meant to generate three types of information: (i) firm-specific characteristics; (ii) the profile of the investment climate in which the firm operates; and (iii) the productivity level of the firm. The logic is that, by controlling for firm-specific characteristics, the data generated from surveys can effectively be used to assess the isolated influence of climate deficiencies on industry performance. Investment climate profiling provides information on those indicators that can be mapped to performance indicators, whereas productivity data are used to compute performance indicators.

Because of the wealth of information sought and the size of the sample in each country, ranging from 200 to 1500 firms, a limited number of surveys are performed each year (current World Bank target lies at 20 per year), making annual global comparisons of investment climate on the basis of survey data impossible. Generally, survey data are used to rate a given country's investment climate against predetermined global data. For that reason, surveys have to date been used to collect investment climate data for developing and emerging economies in North Africa, SSA, South-east Asia and Latin America. Climate survey statistics are ultimately used in formal assessments of investment climate, and are available online to analysts.

Limitations of conventional frameworks as measures of competitiveness in agro-industries

For appropriate application of the different indices of business climate assessment, it is important to understand the underlying assumptions, the data used and the index computation methods. The discussion of business environment assessments presented in this section reveals some cross-cutting features of the existing methods, and the strengths and limitations of these methods in assessing agro-industry competitiveness are summarized in Table 1.

The assessment reveals that the business climate is generally described at national level to focus on macro-level determinants of investment attractiveness. As a result, the effect of local variations in access to (or application and enforcement of) the national determinants is not captured. Further, the local business environment is the relevant environment in which the business operates, and is therefore the environment of interest for competitiveness assessment. The BCI comes close to addressing this limitation except that, although analyses are focused on microdeterminants of competitiveness, conclusions are still made at the national level.

Likewise, for industries with unique characteristics, the value chain traits become as important as, if not more important than, broad national ones in determining competitiveness. To the extent that the industry is global, nation-specific descriptions of competitiveness are less important than, say, local value chain coordination and the extent to which these value chains are integrated into the global value chain. In this case, stimulating supply response, strengthening supporting markets and strengthening end-market demand become critical in creating and sustaining competitiveness. We argue that these features are apparent in the agribusiness sector.

In the agribusiness sector, 'climate' issues such as condition and type of downstream markets, proximity to markets, compliance with sanitary and phytosanitary standards, the presence of subsidies in local and foreign markets, prevailing food security policies, rural infrastructure, farmland ownership structures, and geographic and climatic conditions can substantially influence profitability. Proximity to input and output markets, for example, is especially important for the agribusiness industry, given the higher level of perishability and bulkiness of products. In emerging markets, quality of infrastructure such as road networks is important for the same reasons, particularly for the effect on accessibility of farming enterprises that are both input suppliers and consumers of products from the agroprocessing industry. Regarding property rights, emphasis is required on state property leasing rights, or the reform of state ownership of properties such as land, biotechnology institutions or agricultural marketing institutions, which are important when considering downstream industry efficiency for agroprocessing industries. Given these unique characteristics, a case can be made for a specialized method of describing the competitive environment for agribusiness firms.

The Nature of Enabling Environments for Agro-industries

Essential factors underlying agro-industry competitiveness in developing countries

One of the most fundamental issues a government must address in formulating policies in a global economy is to define its own role in fostering economic progress. The role of the state, at its most basic level, calls for the provision of laws that define property rights, enforce contracts and resolve disputes. In this

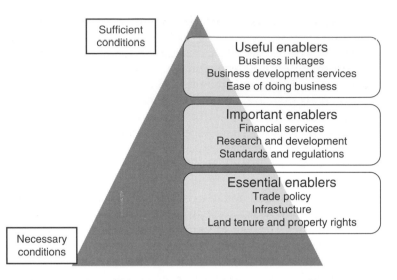

Figure 3. Hierarchy of enabling needs for agro-industry competitiveness.

sense, without state action, markets could not exist. Governments can play an even larger role by investing in infrastructure that contributes to the efficient functioning of markets. In Figure 3, we identify a hierarchy of enabling needs that a government can consider in addressing its role in advancing economic progress. The nine enablers were derived from proceedings of two FAO regional workshops (Eastern Europe and Latin America) on 'Comparative Appraisals of Enabling Environments'. The proposed hierarchy divides state actions into three levels of activities that characterize and assess enabling environments for agro-industrial enterprises. At the base of our pyramid, the state must provide *essential enablers* that will make possible the function of markets and enterprises. In this category we place items such as rule of law (e.g. contract enforcement, property rights), provision of infrastructure and a conducive trade policy. So-called *important enablers* are second-order activities that the state can and often does provide, such as finance, transportation and information. Finally, we define *useful enablers* as sufficient but not necessary conditions to include grades and standards, linking small farmers to formal markets and business development services. In the following section, we discuss each of the enablers starting at the bottom of the pyramid with essential enablers and moving up to important and useful enablers.

Essential enablers

Land tenure and property rights
Land is one of the most important assets for people throughout the world. It constitutes the most significant part of the assets base for the rural and urban poor – especially in developing countries where farming is the main source of livelihood. Secure land tenure and property rights are as central to peace and

stability as are rule of law, good government and economic development. From an economic standpoint, land is a 'keystone factor' of production for economic activities. Traditional economic theory assumes exclusive, transferable and enforceable property rights in land when considering it as a factor of production. Given these assumptions, resource endowments, preferences and technology are sufficient pillars of the traditional economic theory to take into consideration. In developing countries, where these assumptions do not hold, omitting institutions, the fourth pillar, from the economic analyses could be misleading. In this case, institutional arrangements, in particular property rights, need to be specified.

Property rights are an essential class of institutional arrangement. Property as a social institution implies a system of relations between individuals. It involves rights, duties, powers, privileges, etc., of certain kinds. Thus, property rights define the use, control and transfer of assets, including land; which of them are lawfully viewed as exclusive; and who has these exclusive rights. Property rights also include enforcement mechanisms to resolve disputes and defend rights. Quality of law enforcement is more important than mere existence of laws. Property rights may also have both temporal and spatial dimensions. There are four basic categories of property rights in land: open access, communal property, private property and state property. Under open access, no exclusive rights are assigned, normally resulting in land degradation. In the case of communal property, exclusive rights are assigned to a group of individuals, whereas under state property the public sector is responsible for managing the land. Finally, under private property, an individual is assigned exclusive rights (Feder and Feeny, 1991).

Systems of ownership rights in land have effects on incentives to use land efficiently and to invest in land conservation and improvement. A robust land ownership system creates powerful incentives for value addition on land, especially where land is scarce or contestable. Establishment and enforcement of these systems, however, are not cost-free. Legal procedures that define property rights and enforcement mechanisms may be very complex and require various types of documents and affidavits, which increase transaction costs. These transaction costs may offset the benefits from enhanced property rights when land is abundant. As theory and empirical evidence suggest, however, when land becomes scarce or technological changes create new investment opportunities, the provision of ownership rights and enforcement mechanisms has the ability to enhance land productivity.

In developing countries, where land may be used as collateral, asymmetric information and uncertainty also play a central role in both formal and informal credit markets. Improving transparency and information on land ownership rights, in particular transfer rights, reduces uncertainty, risk aversion and the 'moral hazard' problems typically associated with rural credit markets. Availability of land title and institutional mechanisms to resolve disputes affects the willingness of creditors to loan when land is used as collateral. Furthermore, existence of formal procedures for registering liens in land titles represents an important enforcement mechanism that provides additional incentives to creditors to make loans. A well-functioning land ownership system also facilitates risk taking and innovation. This tends to be very important in rural areas, where people tend to be conservative and risk averse. Thus, systems of ownership rights that increase incentives for efficient use of land also stimulate a more efficient credit market (Feder and Feeny, 1991). As mentioned earlier, institutional arrangements to put in place systems of property rights are not cost-free. Transaction costs arising from these institutional arrangements may well offset their benefits and exclude the poor.

Infrastructure

Since the mid-1980s, evidence of the increasing concern and debate about the impact of infrastructure on economic development has surfaced in a large number of empirical studies. Infrastructure is defined to include the sectors of transportation, water and sanitation, electric power, communications and irrigation. Those sectors represent a large portfolio of expenditures in most countries, with a range from one-third to one-half of public investment, or 3–6% of GDP. Much of the formal research on the effects of infrastructure has examined macroeconomic or industry-wide variables. Usually, they all find that infrastructure has a significant and positive effect on economic growth. More recent studies have approached this empirical question by using micro-level data, avoiding the problem with externalities.

The positive impacts from infrastructure are not derived from investments in physical facilities, but rather from the services generated. Four conditions are necessary to realize these impacts on economic development:

1. The basic macroeconomic climate should be conducive to an efficient allocation of resources.
2. Infrastructure projects can raise the returns to other resources only when a sufficient complement of other resources exists; infrastructure investments cannot create economic potential, only develop it.
3. Infrastructure activities that have the most significant and durable benefits in terms of production and consumption are those that provide the degree of reliability and quality of services desired by users.
4. Infrastructure is more likely to be economically efficient, and to have favourable impacts on the environment, when it is subject to user charges.

The impact on international competitiveness is quite direct. Inadequate infrastructure cripples the ability of countries and industries to engage in international trade. Increased globalization has resulted not only from economic factors such as trade policy and the integration of financial markets, but also from major advances in communication, information technologies and transportation. Those infrastructure investments are linked to productivity and to aggregate sales (Peters, 1992).

Trade policies

Trade policies play a critical role in determining industrial competitiveness through two main avenues: first, directly, through the impact on the cost of production and the price of commodities and products; and second, indirectly, through the impact on market access and on global market trends. The first of these routes is, to some extent, within the control of national policy makers, who can strategically exploit policy instruments to direct creation of competitive advantages or to reinforce existing ones. The second is within the control of only those nations with enough global market power in specific industries, or, more commonly, is subject to the lobbying powers of those market players who control the largest global market share. In either case, trade policies may facilitate increased firm-level productivity; they may also severely stunt industry growth.

Conventional assessments of the business environment address trade policies to a limited extent, often focusing narrowly on the impact of tariffs and costs of trade on business profitability. In practice, however, trade control instruments may take numerous forms, ranging from direct taxes and quantity controls, to indirect instruments such as monetary policy and technical measures. In the agribusiness sector, the debate is further complicated by the presence of sector-specific characteristics with respect to market organization and trade patterns. We discuss here sector traits pertaining to market access, market power and preferential trade agreements. As the discussion shows, these unique characteristics of the agriculture sector call for specialized analysis methods in business environment assessment.

Market access regulation, a major globalization challenge of this decade, has been shown to have, in specific cases, significant impacts on industry profits and growth. In the agriculture sector, non-economic (sometimes irrational) motivations for trade protection are fairly high, access of agricultural goods into major world markets is severely restricted by the presence of subsidies and world market prices are substantially distorted. In some cases, legitimate trade-monitoring measures (e.g. sanitary and phytosanitary standards) have been used to restrict entry. Market access has also been restricted through tariff escalation, an instrument employed to protect or develop one's manufacturing industry using inexpensive raw material imports. Those forms of trade barriers in local or foreign markets are important in explaining international differences in competitiveness of specific subsectors.

Government intervention in agricultural commodity marketing has historically been quite high, evidenced by the presence of state-owned commodity marketing boards and various forms of price controls (direct controls, subsidies and market supply restrictions). The agriculture sector is one of the few sectors in which producer associations are not only legal in most countries, but also prevalent and sometimes encouraged. Export boards with autonomous export authority also exist for specific products in both developed and developing economies. As a result, global and local agricultural markets are distorted by the presence of big players, with implications on business competitiveness. However, this trend or ability is being dramatically curtailed.

The agriculture sector, apart from other natural resource sectors, is a large player in north–south preferential trade arrangements (PTAs). Notwithstanding the potential openness gains, these trade arrangements can also distort market incentives. In many cases, PTAs develop demand for specific commodities, hence foster competitiveness and industrial growth for those commodity sectors. Assessment of agribusiness climates, therefore, ought to highlight such trading arrangements.

Policies, tariffs and quotas for imported products

Trade and domestic support policies related to import of competing products are areas of contention between many countries, as seen in the World Trade Organization. Many WTO member states argue in favour of some reductions in trade restrictions and domestic support, but there is little consensus on the rate of reduction, commodities to be affected and special and differential treatment. The distortions caused by these policies potentially hamper the ability of agribusinesses in emerging markets to be as productive and as profitable as possible.

Considering the years of policy analysis conducted by OECD and the World Bank, developed economies have had relatively high tariffs and domestic support for agricultural products relative to developing economies. As the recent case of the USA versus Brazil suggests, at least in cotton markets, these policies have put emerging economies like Brazil's at a disadvantage in international markets.

A concern about global competition in light of agricultural policies is that the competition is not fair. Since some countries, especially developed countries, are using trade and domestic policies to protect and/or promote their products, developing countries should do the same. The policies of developed countries may limit the opportunities of emerging economies. Retaliatory trade or domestic support policies, however, are not an ideal response to policies of other countries. Given the budgets of many large, developed countries, the outcome of policy competition is that smaller developing countries will be unable to compete. Instead, the Brazilian example exemplifies an approach that developing countries should consider: if the policies of other countries are unfair, challenge them in the appropriate dispute settlement bodies. As recent events have shown, developing countries can use the WTO and other international bodies to their advantage to bring about change.

'If you can't beat them, join them' is an adage that should not be followed as it relates to agricultural policy and trade. One way for developing countries to 'join them' is by using policies to promote domestic industries under the guise of promoting an infant industry. The old infant industry idea is simply that – old and, worse, debilitating. The infant industry argument suggests that countries could be competitive in a particular industry if only the countries were able to provide appropriate protection through trade and domestic support policies so that the infant can mature. The problem with the infant industry idea is that once policies are in place they are hard to remove. The infant rarely is considered to have grown old enough for protections to be removed. Like a child who is never allowed to mature by an overly protective parent, the infant industry struggles to become efficient and competitive in global markets because the protection of the government prevents maturation. Infant industry policies often promote and foster inefficiencies. If the infant industry produces an input for a domestic agribusiness, the inefficiencies, often seen as increased costs, will transfer to the agribusiness. For governments to create an enabling environment, they must create policies that promote efficiency through investment with technology transfers. Internal investments are longer-term goals, but in time they will generate great benefits. Competition will promote great efficiencies, which will promote industries that thrive.

Important enablers

Norms, standards, regulations and services related to production
Among the greatest benefits and challenges of globalization is the meeting of different cultures. One expression of this meeting of cultures is the differences of opinions as they relate to norms, standards, regulations and services related to production, processing and distribution of agrifood products. Because of history, perception of national identity, religion, etc., citizens of different countries have different conceptions of food and its role in their lives. For example, Europeans who want traceability of food products back to the farm of origin have been deeply influenced by food scares. That interest is not different from that of the US producer who produces genetically modified soybeans and believes that, because science has generated this product and the US government has approved its use, GM soybeans are safe and beneficial. Consumers and producers in other regions of the world do not necessarily share those perspectives. In particular, some nations' approach to animal welfare and GM products are adamantly opposed by others. These generalizations point to the varying views that consumers have of norms and standards.

Diversity of perspectives provides a complex environment for agro-enterprises. If all of the norms and standards were on a single continuum of relative restrictiveness, an agro-enterprise could simply produce at the higher standard and sell products to all at that higher standard. But standards and norms may not fall on a single continuum of restrictiveness. Consider an agrifood firm that distributes nationally and exports to two different countries. One country prescribes a maximum residue level that the firm's product must satisfy. Another country demands that the agribusiness firm assure that the product is produced in a manner deemed equitable, as determined by the grocery store purchasing the product. Finally, on the domestic market, the agribusiness firm must achieve a high quality standard because local consumers are familiar with the product and are particular about its quality. The first problem is how to manage different standards for different customers. The achievement of one standard may be in conflict with the achievement of the others, at least in the short run. To sell the same product to all three markets may be costly, because a market may not be interested in the standards of the other markets. Above all, the cost of achieving all of the norms and standards of the three different markets may be prohibitive.

Countries change standards. Agro-enterprises must keep abreast of the various quality and hygiene standard changes. According to the WTO's Sanitary and Phytosanitary Agreement, countries are obliged to notify exporters of changing standards that fall under the purview of that agreement via enquiry points. However, every norm and standard do not fall under the purview of the Sanitary and Phytosanitary Agreement. Additionally, firms face not only the standards of the countries that import the products, but also the standards imposed by importing firms, especially food retailers. In this complex web of standards, governments can create enabling environments to assist agricultural producers and other agrifood chain stakeholders, by providing any information that can keep firms aware of changing standards. Governments can also provide financial and technical assistance to meet the standards of importers; in addition to supporting research institutes and product marketing institutes. Using the mechanisms of the WTO and other international organizations, governments can access resources to help firms meet new standards.

Research and development

Agricultural research has long been recognized as critical to increasing agricultural productivity and thereby reducing extreme poverty and hunger (Ruttan, 1975; Herdt, 2009). Equally important but less accredited is the role of agricultural research in establishing and maintaining the competitiveness of the agro-industrial sector. Numerous examples showcase how technology can reverse a competitive advantage bestowed by nature. For example, Israel's agricultural exports (currently valued at approximately US$600 million annually) continue to grow despite the country's near-desert conditions. Technologies for most non-agricultural industries may be transferred between countries – usually from developed to developing countries – with minimal or no adjustments, thereby allowing for the much-acclaimed 'technological leapfrog' by developing countries. For agro-industries, however, culturally specific food consumption patterns, coupled with diverse agro-ecological conditions, may limit the scope of technology transfer. This specificity underscores the importance of agricultural research in creating an enabling environment for agro-industries.

The structure and context of agricultural research have changed significantly over the last 2 decades. Table 2 shows the annual growth rate of global public agricultural research expenditures from 1976 to 1996. Three key trends are worth pointing out. First, the growth rate of agricultural research expenditures has declined globally in both developing and developed countries since the mid-1970s. Second, while developing countries' growth rates were, on average, higher than those of developed countries, the Asia and Pacific region had the highest growth rates followed by the Middle East and North

Table 2. Annual growth rate (%) of global public agricultural research expenditures, 1976–1996. (From Pardey and Beintema, 2001.)

Region	1976–1981	1981–1986	1986–1991	1991–1996	1976–1996
Developing countries	7.0	3.9	3.9	3.6	4.5
Sub-Saharan Africa	1.7	1.4	0.5	−0.2	1.5
China	7.8	8.9	2.8	5.5	5.2
Asia and Pacific, excluding China	8.2	5.1	7.5	4.4	6.5
Latin America and the Caribbean	9.5	0.5	0.4	2.9	2.5
Middle East and North Africa	7.4	4.0	4.2	3.5	4.8
Developed countries	2.5	1.9	2.2	0.2	1.9
Total	4.5	2.9	3.0	2.0	3.2

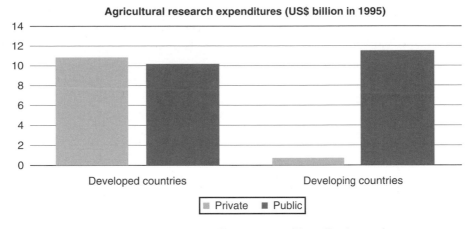

Figure 4. Agricultural research expenditures, 1995. (From Pardey and Beintema, 2001.)

Africa. Third, SSA countries had the poorest performance, with an average annual growth rate of only 1.5% for the period of analysis and actual decline in the early 1990s.

Agricultural research expenditures by the private sector are large and growing, especially in developed countries. Figure 4 shows the breakdown of agricultural research expenditures by funding sources in 1995. For developed countries, private-sector expenditure in agricultural research (US$10.8 billion) has outstripped public expenditure (US$10.2 billion), whereas in developing countries agricultural research is still very much the domain of the public sector (US$11.5 billion in the public sector versus US$0.7 billion in the private sector).

Given the prominence of public funding and public research institutions (mostly national agricultural research organizations (NAROs) and the Consultative Group of International Agricultural Research (CGIAR)) in developing countries, emphasis should be placed on bridging the gap between research and commercialization in order to enhance competitiveness of the entire agro-industry value chain. As currently structured, a distinct disconnect exists in agro-industry value chains for most developing countries. This is the gap between product development (undertaken mostly by public research institutions) and commercialization (mostly by private agribusiness firms). Numerous technologies have been developed through publicly funded NAROs and CGIAR research centres, such as the International Maize and Wheat Improvement Center, International Crops Research Institute for the Semi-arid Tropics, International Institute of Tropical Agriculture and International Potato Center. Most of these technologies are still 'sitting on the shelves' while the private sector is struggling to commercialize a limited range of outdated technologies.

If agricultural research is to enhance agro-industry competitiveness in developing countries, policy makers should focus on identifying ways to better coordinate the flow of agricultural technology from public discovery to private use. In accomplishing this objective some important questions should be addressed:

- What are the key barriers (cost- and non-cost-related) to technology transfer between public research institutions and private seed companies?
- What are the sources of institutional innovation in promoting successful public–private partnerships?
- What is the role of the policy and regulatory framework in impeding or accelerating technology development and transfer?

Financial services for agro-industries

Access to finance is one of the key constraints to agro-industrial development and success. Firms in the agricultural sector often have difficulty in accessing capital for either new ventures or expansion of existing business as they are perceived to be high-risk[1] businesses with low returns. As a result, an agribusiness portfolio is not an attractive option for investors who tend to have an appetite for high returns if high risk (and a narrow inter-temporal investment horizon) is involved. Importantly, the risk profile of agribusinesses differs from that of other sectors in that the former are faced with both inter- and intra-marketing year price and production risks. These risks have further been propelled by the increased globalization of the free trade in agricultural commodities. The agribusiness credit crunch is further exacerbated by most bankers' ignorance of the sector, which increases the chance of loan applications being dismissed entirely on the perception of low profitability. To this end, there exists a dire need to create an enabling environment for the provision of financial services to the sector.

Though prevalent in developing countries, the problem of accessing finance is not confined to these countries as it was an issue at some point for industrialized countries as well. For instance, concerned with farm credit, the US government established a farm credit system in 1916, which is currently administered by the Farm Service Agency of the US Department of Agriculture. The extensions of this programme and supportive legislature have been fundamental to the development of the US agro-industry.

The intangible nature of financial services requires a strong regulatory environment for financial markets to develop and function effectively. As a result, in the short term governments need to take the lead in building confidence, trust and stability among participants in these markets. One way of doing so would be to act as guarantor for loans to the agro-industry. In the long term, well-defined property rights, particularly in farming, would be crucial to enable use of real estate as collateral in accessing traditional financial markets. It may be necessary to build a farm credit system following the US model. Overall, the evolution of financial markets for agriculture has been observed to parallel general financial market development; hence a necessary condition for the former is functionality of the latter.

---

[1] Risks inherent in agribusiness include changes in climate (drought, flooding), price volatility and production variability.

Since agro-industries are a high-risk but relatively low-margin segment of the economy, their success will require innovative and flexible ways of hedging against risk. One means of reducing price risk would be through the use of commodity futures exchange markets. Proper functioning of a futures market depends on enforceability of contracts and a dependable information system. Crop insurance, on the other hand, would be instrumental in mitigating production risks due to natural catastrophes.

While creating an enabling environment for agribusiness finance, special attention needs to be accorded to small- to medium-sized agribusiness entities considered too small to access traditional capital markets, but too large to depend entirely on personal or family savings. Agribusiness entities in this size category are increasingly becoming important to developing country governments ever mindful of the need to ensure food security for their populations.

Useful enablers

Ease of Doing Business

One of the functions of governments is to regulate economic activities to reduce inefficiencies arising from market failures, in order to improve economic and social outcomes of specific countries. Regulations, however, have to be done in less costly and burdensome ways to facilitate doing business and to attract investments that promote economic development and ultimately reduce poverty. Governments around the world have implemented macroeconomic reforms to attract foreign capital through the domestic and international private sectors, thereby making domestic industries more competitive. Since 2004, those governments' reforms have been considerably influenced by the World Bank's Doing Business project, because countries want to improve their rank in the Ease of Doing Business Index so as to provide signals to investors interested in FDI. In spite of the implementation of those reforms, vibrant private-sector engagement in specific economies remains limited, poverty rates are high and growth continues to be static in a number of countries.

Although macroeconomic policies are undoubtedly important to promote economic development, it is now widely recognized that the quality of business regulation and institutional arrangements to enforce it are determinants of economic prosperity. Findings of Ease of Doing Business show that a hypothetical improvement of all aspects of the Doing Business indicators to a level commensurate with the top quartile of countries is associated with an estimated 1.4–2.2 percentage points in annual economic growth, whereas a similar improvement in macroeconomic and education indicators results in 0.4–1.0 percentage points in growth. Ease of Doing Business consistently indicates that countries with excessive regulation for doing business and weak property rights generally have lower labour productivity, higher poverty rate, more exclusion of the poor from doing business, slower economic growth rate, lower human development indicators and higher levels in the incidence of corruption. Countries would benefit from simplifying their regulations for doing business in two ways. First, entrepreneurs not only would spend less time dealing with government regulations (e.g. business licensing and registering, contract enforcement and resolving disputes), but would also focus their energies on producing and marketing their goods and services. Second, governments would spend fewer resources regulating and more providing basic public goods such as infrastructure to improve economic and social outcomes.

The Doing Business project (World Bank, 2007) reports that Sweden, a top-10 country in Ease of Doing Business, spends US$7 billion a year, or 8% of the government budget, and employs an estimated 100,000 government officials to deal with business regulations. If Sweden cuts expenditures on the administrative burden by 15%, the savings would amount to 1.2–1.8% of the total government expenditures, or approximately half of the public health budget. The data analysed in Doing Business highlight that reducing the number of procedures to only those truly necessary (statistical, tax and social security registration), abolishing the minimum capital requirement and using the latest technology to make the registration process electronic have produced excellent results in Canada, Honduras, Mexico, Pakistan and Vietnam. Those reforms may be difficult to implement because political will in government and private sectors may differ, but they have beneficial effects beyond business entry. Cross-country evidence has shown that the countries that have the greatest need for entrepreneurs (to create jobs and to promote growth) – poor countries – put the most regulatory obstacles in their way.

Business development services

Successful investments in small- to medium-sized enterprises must be paired with appropriate firm/business management assistance and access to value-added business networks in emerging markets. Although investors worldwide are pursuing investments in emerging markets, studies have confirmed those efforts to be particularly challenging.

Business development services draw upon formal qualification in areas such as finance, accounting, marketing management, economics, law and other technical expertise. Aside from an academic grounding in one or more of these disciplines, however, possibly a more essential prerequisite, and one to which the industry refers, is experiential knowledge. As the term suggests, it is the knowledge obtained from actual experience ('learning by doing'). How one obtains such knowledge, e.g. through structured mentoring by and/or informal consultation with more seasoned professionals, is especially important since many services do not require prior formal qualification in a 'closed' profession such as accounting, engineering, law or architecture. Finally, provision of these services will often draw, but only selectively and for limited periods, on particular domain expertise and knowledge. For example, vetting of a proposed investment in an agribusiness company will require expert 'technical knowledge' of the product in question, including issues such as plant breeding, processing, packaging and certification, among others.

Formal academic and professional qualifications are often, but not always, a necessary condition for undertaking a particular service. More important, especially for undertaking critical services within the investment cycle, is experiential knowledge, often confused with domain expertise. The latter refers to expert knowledge, which, as noted earlier, may be needed to provide various services associated with the investment cycle. One such example is technical knowledge of processes for producing high-quality food products. Clearly, it is needed to evaluate proposed investment in a plant producing such a product. What proved equally significant in terms of appraising the investment was the supervisors' and managers' accumulated experience, through past jobs, in operating and maintaining the equipment in question, as well as other processing-related activities.

The international donor community and national governments have attempted to strengthen business management services, especially those provided by local professionals. Broadly speaking, their efforts fall into three categories: (i) increasing the supply of providers; (ii) stimulating the demand for various services; and (iii) addressing issues of both supply and demand within the parameters set by specific investments.

Donors and state agencies have promoted small- and medium-sized enterprises in three different areas: financial services, business development services and government-mediated business environment. An immediate motivation is the desire for a higher degree of self-financing because of cutbacks in support from government and private givers. This objective, however, also coincides with growing realization of the need for commercial acumen and expertise to fulfil their corporate aim of improving the lives of poor and disadvantaged people in the countries in which they are operating. Indeed, some have been severely criticized for past, ill-informed interventions that, far from improving the lot of poor farmers, have actually worsened it. Overall, there is increased realization that a market-based approach is much more likely to improve the living standards of those people the organization is seeking to assist.

Business linkages

Linkages for large agribusinesses in the supply chain are both horizontal (i.e. between enterprises that are on the same level of the supply chain) and vertical (i.e. between enterprises that are on different levels of the supply chain). Most horizontal linkages pertain to large and small agro-industries, while vertical linkages pertain to large agro-industries, farmer groups and buyer networks. Of the two, horizontal linkages are less common due to the lack of incentives for large agro-industries to pursue such relationships. Large industries may subcontract to their smaller counterparts in order to satisfy a market opportunity. Such arrangements may not have indirect spillover effects like the transfer of technology and information. Alternatively, large agro-industries may jointly bid for contracts with smaller firms and, in so doing, increase their access to markets.

On the other hand, vertical linkages are more beneficial to both parties (large agribusinesses and farmer groups), because in most cases they entail long-term direct and indirect benefits. Large agro-industries pursue linkages with farmers or farmer groups in order to access a steady supply of raw materials. The extent of these linkages may vary from a 'one-time' engagement to a long-term contractual relationship. In the latter case, the large firm may invest in training farmers in the production of a particular agricultural commodity; further, it may also provide capital for the purchase of agricultural inputs. The goal of such a relationship is to ensure a steady supply of a desired or specified product for the large firm. Farmers, in turn, derive benefit from this relationship. First and foremost, they now have a steady market for their farm output, and therefore a steady income stream. In addition, they have acquired farm management skills that enable them to produce a quality product. This is to farmers' advantage – should the business relationship with the large firm cease, farmers are competitive enough to pursue relationships with other firms. Further, farmers now have access to technologies in the form of improved inputs like seeds and fertilizers that should improve their productivity and profitability.

Vertical linkages should be encouraged by public policy makers because they have the potential to eradicate poverty at the smallholder farm level through income generation. Specifically, policy support should be directed towards the formation of farmer groups so as to reduce the risk of insufficient supply. In addition, farmer groups are better equipped than individual farmers to negotiate favourable contract terms with large agribusinesses. The business relationship is economically attractive only if farmers realize higher profits than they would have had they pursued alternative markets.

Reforming Enabling Environments

How can we reform the public process, in the context of radical change, to creatively enact policies that improve the relative competitiveness of agro-industries in emerging markets?

As discussed in the literature, 'economic reform' is the process by which emerging economies are transformed from state-led to market-driven principles with the goal of advancing economic prosperity. This view puts the government as the primary driver of the reform process with measures that, in sequence, include privatization, tax reform, fiscal discipline, trade liberalization, deregulation of economic activities, price liberalization, decontrol of interest and exchange rates, elimination of state subsidies and enforcement of intellectual property rights. Because of the rapidity of globalization, however, we observe a more parallel reform process, one in which state and private sectors act in concert to create an enabling environment. Therefore, in this chapter we take a more 'nuanced view' of the reform process, capturing a more dynamic (i.e. one that goes through multiple phases) and inclusive process in which the private sector can play a leading role in certain cases. Further, we acknowledge the wide diversity of experiences in transitional economies over the last 2 decades emanating from different starting points, economic structures, the conduct of both private and public stakeholders and the desired outcomes. The central role of the agricultural sector, coupled with a long history of state control, places agricultural market reform at the centre of most economic liberalization efforts in transitional economies. For brevity, we focus on agricultural sector reform, although of course the agricultural sector exists and evolves in a more dynamic and broader macroeconomic environment.

Like most other private enterprises, agricultural firms tend to be risk averse. Consequently, they require a higher rate of return (a risk premium)

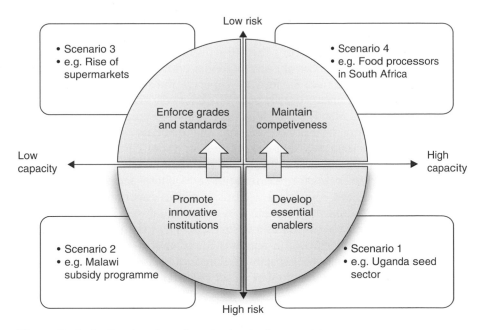

Figure 5. Agricultural sector reform in developing countries.

before they can invest in a risky environment. From the perspective of a private agribusiness enterprise seeking to maximize expected profits, the enablers discussed in the preceding section can be seen as ways to augment the expected profit function by altering the production functions, the input and output prices, and lowering the risk and uncertainty.[2] A key distinguishing feature of the agricultural sector, compared with other sectors of the economy, is the high prevalence of risk and uncertainty that result from both natural and human-made phenomena. While some sources of risk are common to all businesses, others are unique to agribusinesses.

Governments in most transition economies have played a key role in shaping the process and outcomes of economic reform, but, in reality, they have varying degrees of 'capacity' that limits their ability to influence the environment in which private enterprise operates. Despite the premise of 'rolling back state involvement' as the basis for agricultural reform, we recognize the essential role of the state in advancing essential enablers, important enablers and useful enablers that target agro-industries.

Presented in Figure 5, our model for analysing the agricultural sector reform process in developing countries is centred on two key variables, risk and capacity. We use those variables to formulate a matrix. Along the vertical axis

---

[2] Frank H. Knight (1921) was the first to distinguish between 'risk' and 'uncertainty'. Risk refers to situations in which the decision maker can assign mathematical probabilities to the randomness of an event based on past experiences. In contrast, uncertainty refers to situations when this randomness 'cannot' be expressed in terms of specific mathematical probabilities.

we map the level of risk and uncertainty agricultural enterprises face when conducting business (investing, starting and expanding agribusinesses). This risk variable summarizes the attractiveness of the 'business climate' from the point of view of current and potential agribusiness investors. Along the horizontal axis we map the level of 'capacity' the state has in shaping the environment for business, usually but not exclusively exerted through the ministry of agriculture. Many of the elements required to create an enabling environment for agro-industries are outside the usual mandate of many ministries of agriculture. Realizing this limitation allows for more careful scenario planning that takes into account the unique context facing each country. From this matrix, four stylized cases for reforming the agricultural sector emerge. In the bottom right quadrant, scenario 1, we encounter an environment that is risky and has high state capacity. Scenario 2 in the lower left quadrant describes an environment that is risky and has low state capacity (promote innovative institutions). Continuing clockwise, in scenario 3 we have an environment that is low risk and has low state capacity (enforce grades and standards) and finally in scenario 4 the environment is low risk and has high state capacity (maintain competitiveness). Below, we discuss each of these four 'stylized scenarios' using country- and sector-specific examples.

Scenario 1: High risk and high capacity
If the state has high capacity and the economy is uncertain, then the state encourages private-sector investment by clearly defining the 'rules of the game'. Such rules would include all 'essential enablers' identified in the hierarchy of enablers pyramid (see Figure 3). A good example of a successful agro-industry under high control and high risk is Uganda's vibrant seed industry (see Box 2). In 1994, the government led a significant transition from a seed system monopolized by the government at all levels (research, production, processing, distribution and extension services) and made way for the private sector by enacting the Agricultural Seeds and Plant Statute (Muhhuku, 2002). The statute provided for 'the promotion, regulation and control of plant breeding and variety release, multiplication, conditioning, marketing, importing and quality assurance of seeds and other planting materials' through the National Seed Industry Authority, the National Seed Certification Services and the Variety Release Committee (Government of Uganda, 1994). Currently, Uganda has a highly competitive private-sector-led seed industry in which local, regional and multinational seed companies produce and distribute seeds in both domestic and East African markets (mostly Sudan and Tanzania).

Scenario 2: High risk and low capacity
Perhaps the least attractive investment climate for agro-industries is the one characterized by limited government capacity and high risk. This scenario applies to many of the least developed countries. In most cases the state lacks the resources or commitment to create the essential enablers. Policies and regulations are ill defined and/or unenforced. If there is strong comparative advantage based on natural endowments, however, some agro-industries have

---

Box 2. Seed industry reform in Uganda.

During the past 5 years the Government of Uganda has launched a number of important programmes, at the core of which agriculture is given high priority. Most notable is the 2000 Poverty Eradication Action Plan, which includes the Plan for the Modernization of Agriculture (PMA). Under the PMA, the government is focusing on the National Agricultural Advisory Services (research and technology development, education, marketing and agroprocessing and natural resource management), which aims to develop an agricultural service delivery system that is client-oriented, i.e. where farmers can determine the services they need. The government also recognizes that an important factor in improving agricultural productivity and production is the use of modern inputs, including seeds.

Uganda has made important progress towards economic growth and poverty reduction since the late 1980s. In the 1990s, annual gross domestic product (GDP) growth climbed steadily to 6.9% from a 3% per annum growth rate in the 1980s. This impressive growth appears to be linked to the set of economic policies that advanced structural adjustments within the economy. The government liberalization of the foreign exchange rate was a significant economic reform that provided incentives to major sectors of the economy: agriculture, industry, trade and tourism. Despite the economic growth record, concerns about the macroeconomic environment still exist. According to an International Center for Soil Fertility and Agricultural Development report (IFDC, 2003), the economic environment remains unfriendly because of the continuous depreciation of the Ugandan shilling, high interest rates and limited access to finance.

In line with the overall government policy, the government of Uganda fully liberalized the seed industry in the early 1990s, and privatized its operations in 1993. By 2005, six new private seed companies had entered the market. In addition to Uganda Seed Project, the private seed companies operating in the country are East African Seed Company, Farm Inputs Care Center, Harvest Farm Seeds, Kenya Seeds, Nalweyo Seed Company Ltd and Victoria Seeds Ltd. Seed Co. International from Zimbabwe and Pannar Seed Company from South Africa have also had their seeds tested and adopted in Uganda through distributors.

The National Agricultural Research Organization (NARO) is responsible for the production of both breeder seed and foundation seed. Currently, agricultural research is fragmented. The research facilities are financially constrained and as a result the operation of the breeding programme is affected. In a competitive, liberalized market, seed companies compete on the strength of their breeding programmes and their ability to introduce new and improved varieties. The relationship between NARO and the seed companies is evolving.

The regional harmonization of seed laws and regulations for East African countries (Kenya, Tanzania and Uganda) that is currently under way has opened national borders for regional seed trade. For private seed companies, the harmonization will present new opportunities (expanded markets) and challenges (increased competition).

---

performed well under these environments. The case of gum arabic[3] from Sudan is a stark example. World trade in gum arabic was estimated to be US$90 million in 2000, with 56% of the traded volume coming from Sudan, and much of the remainder from Chad and Nigeria (Cecil, 2005). While these countries possess the natural resource, their governments have done little to create an enabling environment for the gum arabic industry.

---

[3] Gum arabic, a natural gum also called gum acacia, is a substance that is taken from two sub-Saharan species of the acacia tree, *Acacia senegal* and *A. seyal*. It is used primarily in the food industry as a stabilizer and is a key ingredient in soft drinks (Fennema, 1996).

If the state has limited capacity, and the economy is uncertain, then the state should provide incentives for creative institutional innovations, like private–public partnerships, civil society organizations, non-governmental organizations and perhaps corporate social responsibility (Box 3).

Box 3. Agricultural input supply programme in Malawi.

Agriculture is the mainstay of the economy of Malawi, where 90% of the population lives in rural areas and more than 70% heavily depends on farming for livelihood. Agriculture contributes about 38% of the gross domestic product (GDP), accounts for 80% of export earnings and employs 80% of the country's workforce. Maize is the dominant staple food, while tobacco, sugar, coffee and tea are important cash crops. Smallholder farmers cultivate nearly 80% of the total cultivated area in the country and devote 85% of their cultivated land to maize production. Agriculture is characterized by low use of purchased-input technologies, high land pressure due to an increasing population growth, periodic flooding and droughts, and continuous cropping of the same land area with little replenishment of the nutrients, leading to decreasing soil fertility. Adoption of higher-yielding seed varieties and fertilizers, better crop management and better water control are critical for improving agricultural productivity in the country.

The Government of Malawi (GOM), as other governments in southern Africa, has implemented a number of input interventions to increase agricultural productivity and food security to ultimately reduce poverty. Major agricultural input interventions include the Starter Pack Program (SPP), the Targeted Input Program (TIP) and the Fertilizer Subsidy Program (FSP). GOM, the European Union and the Department for International Development (DFID) among other donors have been jointly funding these agricultural input interventions.

In 1998, the GOM initiated a 'free' SPP aimed at stimulating use of fertilizers and improved seeds of cereals and legumes to raise national output to reduce food imports. In addition to supporting extension services, under the auspices of the SPP, fertilizers and improved seeds of cereals and legumes were supplied free-of-charge to smallholder farmers. The SPP had a country-wide coverage and resulted in a significant increase in agricultural production, reducing production deficits. Its financial and economic viability, however, has never been comprehensively evaluated. Thus, the effectiveness of the SPP has been questioned.

In response to extreme drought and low agricultural productivity in early 2005, the GOM introduced a universal FSP in replacement of the TIP during the 2005/06 agricultural season. The FSP favours poor farmers with the land and human resources to use fertilizer efficiently, when they could not otherwise afford to buy enough fertilizer and improved seed. The FSP focuses on maize as a food crop and tobacco as a cash crop. Thus, two subsidy fertilizer packages are distributed country-wide: one for maize and the other for tobacco. Due to large amounts of cross-border trade, vouchers are to be used to effectively target Malawian smallholder farmers as opposed to neighbouring countries' farmers. Vouchers of different colours are issued depending on the fertilizer package to be redeemed. Also, the subsidized price to be paid by the beneficiary is indicated on the voucher to avoid fraud and ambiguity. Maize farmers pay lower subsidized fertilizer price than tobacco farmers. On average, the GOM subsidize about 76% of the commercial price of fertilizers. In addition to fertilizer vouchers, farmers are entitled to receive a maize seed bag for free once they redeem their fertilizer vouchers.

In the 2006/07 agricultural season, almost 1.5 million farmers received vouchers for the purchase of 150,000 t of fertilizers and two million farmers received vouchers for free maize seeds. The Ministry of Agriculture and Food Security allocates over 50% of its budget to

continued

Box 3. Continued.

implementation of the FSP. No comprehensive evaluation has been conducted on the implementation and impact of the FSP, but it has bolstered agricultural output, increasing food security. In combination with favourable rains, the FSP has significantly contributed to increases in maize production. It has also increased smallholder farmers' access to fertilizer and to improved maize seed. Although Malawi was experiencing grain deficits in 2005, in 2006, the country produced an estimated surplus of 1.2 million tonnes of maize. Part of this surplus was exported to Zimbabwe, Swaziland and Lesotho, raising smallholder farmers' income. It is worth noting that these input interventions have considerably reduced farmers' demand for agricultural inputs from the private sector, shrinking private-sector participation in agricultural input and output markets.

Scenario 3: Low risk and low capacity

In scenario 3, the state has relatively little capacity, and the economy has a greater degree of certainty. Due to the low-risk environment, the private sector will emerge to exploit any economic and commercial opportunities available in the agricultural sector. The state should focus on facilitative policy measures that seek to provide public goods. Policies that fall under this category include collection and dissemination of market information, standardization of product grades and their enforcement, standardization of weight and volume measures, and setting and enforcing standards and laws to protect consumers and the environment. Due to the public good dimensions that are associated with these types of services, we envision that they will be provided by government agencies at public expense.

The rise of supermarkets in developing countries (East and South-east Asia, Latin America and a number of countries in SSA) epitomizes this scenario (see Box 4). Some key lessons on reforming the agricultural sector can be drawn from this case study. First, transformation of the agro-industry does not require all enablers to be in place before the private sector can invest. Instead, some push factors in developed countries, coupled with the prospects for higher margins in developing countries, can attract FDI. Second, the sequencing of the reform process (usually framed with government playing the lead and the private sector following) can be reversed (government adapting to meet the needs of the private sector), and in most cases takes place as a complex and dynamic process with checks, balances, negotiations and feedback loops between the private and public sectors. Third, once a sector of the agro-industry builds critical mass, the private sector often evolves to fill in the gaps in the facilitative measures that are not provided by government (e.g. setting up and enforcing grades and standards).

Scenario 4: Low risk and high capacity

In scenario 4, the state experiences improved capacity and the economy has lower risk. This scenario often arises at the end of a successful reform process. The role of government should be limited to measures that maintain and sustain competitiveness through the provision of public goods, enforcement of antitrust regulations and protection of intellectual property rights (Box 5).

Box 4. The rise of supermarkets in transition economies. (Adapted from Reardon *et al.*, 2004.)

The rapid rise of supermarket chains in developing countries has transformed the market structure, participant conduct and economic performance of agrifood systems in Africa, Asia (excluding Japan) and Latin America (Reardon *et al.*, 2004). Intense competition in home markets, coupled with much higher profit margins in developing countries, spurred the flow of foreign direct investment (FDI) from American and European supermarket chains (Gutman, 2002). The liberalization of the retail sector in most transition economies in the 1990s also created new opportunities for investments.

In Latin America, the share of supermarket sales as a percentage of national food retail sales grew from 10–20% in 1990 to about 50–60% at the turn of the millennium (Reardon *et al.*, 2004). The supermarket growth trend in East and South-east Asia has been generally similar to that of Latin America with China topping both size (US$55 billion in 2003) and the growth rate (30–40% annual growth in 2003) (Hu *et al.*, 2004). Central and Eastern Europe (CEE) has seen three phases of penetration by supermarkets. First came northern CEE (Czech Republic, Hungary, Poland and Slovakia), in the mid-1990s, where the share of supermarkets in food retail had risen to 40–50% by 2004. The second phase was in southern CEE (Bulgaria, Croatia, Romania and Slovenia), where the share grew to 25–30% in 2004 and continues to grow rapidly. The third phase was in Eastern Europe, 'where income and urbanization conditions were present for a takeoff but policy reforms lagged, so that the share in Russia, for example, is still only 10%' (Reardon *et al.*, 2004). In SSA, the supermarkets trend is in the early stages, with South Africa and Kenya leading the field and expanding in their respective regions of the continent.

As noted by Reardon *et al.* (2004), supermarket chains in developing countries have been shifting over the past few years away from the old wholesale procurement model towards a new model in order to close the gap between their supplies and their needs. The four key pillars of a new kind of procurement system are: (i) specialized procurement agents, or 'specialized/dedicated wholesalers'; (ii) centralized procurement through distribution centres, as well as regionalization of procurement; (iii) assured and consistent supply through 'preferred suppliers'; and (iv) high-quality and increasingly safe products through private standards imposed on suppliers. The first three pillars require an organizational change in procurement, while the fourth has resulted in institutional change in the agro-industry value chain.

Box 5. Collusion in South Africa's milling industry.

The South African wheat-to-bread value chain comprises farmers, millers, bakers, retailers and consumers in successive stages of value addition. This value chain is currently characterized by high degrees of concentration, especially in the processing (milling and baking) sector. The baking industry is the major client of the wheat milling industry and most of the major millers have vertically integrated with the industrial plant bakeries (National Agricultural Marketing Council, 2003). Illustrative of the relatively high levels of concentration, currently, four main milling companies account for 87% of the total milling capacity in South Africa, while in the baking sector approximately 80% of the bread production is in the hands of six large baking companies or groups. These levels of concentration are primarily a result of measures during the period of regulated marketing that restricted the registration of millers and bakers (National Agricultural Marketing Council, 2003).

continued

Box 5. Continued.

During 1997, the South African wheat-to-bread value chain came under fire because of collusion in fixing the prices of bread in South Africa. During December 2007, independent bread distributors in the Western Cape province of South Africa lodged a complaint with the National Department of Trade and Industry's Competition Commission that bakeries owned by three of the big milling and baking companies had unjustifiably raised prices by between 30c and 35c a loaf a week before Christmas. This behaviour raised suspicion of anticompetitive conduct and the Competition Commission referred the case to the Competition Tribunal for investigation (Seria, 2007). When the Competition Commission referred the case Premier Foods applied for the Competition Commission's leniency programme. Premier Foods provided the Competition Commission with the details of meetings with its competitors, at which national price increases and discounts were discussed. The result was that the company escaped being penalized for contravening the Competition Act (Act 89 of 1998) for anticompetitive behaviour by colluding to fix the price of bread (Crotty, 2007; Seria, 2007).

With the investigation of anticompetitive behaviour under way, Tiger Brands, another big milling and baking company, immediately undertook to cooperate with the Competition Commission. As part of their cooperation Tiger Brands instituted a wider, national and independent investigation into its milling and baking operations by commissioning a commercial law firm, forensic auditors and economic consultants to investigate the matter. Although the independent investigations found no evidence of abnormal pricing or of consumers being adversely affected the investigation did find evidence of meetings between Tiger Brands employees and the employees of competitors (Seria, 2007). These meetings amounted to anticompetitive activity expressly prohibited by the Competition Act (Act 89 of 1998). Tiger Brands consequently admitted that the activities of certain employees had contravened the Competition Act (Act 89 of 1998). These activities amounted to collusion, along with other millers and bakers, to fix the price of bread in the Western Cape. The Competition Commission consequently fined Tiger Brands R98.8m and granted the company leniency against prosecution subject to agreeing to assist the Competition Commission with investigations into possible collusion among millers and bakers (Crotty, 2007). The fine imposed on Tiger Brands represented 5.7% of their bread sales during 2007 (Seria, 2007).

Conclusions

Agro-industries are an engine for growth in rural economies and the agro-industrial sector plays a central role in the economic development of low- and middle-income countries. The rise of global markets based on competitive advantage is, however, increasingly forcing policy makers to make assessments of the 'enabling environment' for agro-industries. After an evaluation of selected measures and indexes of an enabling environment we conclude that standard measures, both macro and micro, are inadequate for evaluating the competitiveness of agro-industries within emerging economies.

Further, we find that agro-industry exhibits unique characteristics that distinguish it from the wider economy, while simultaneously vast segments of the food and fibre markets are becoming well integrated into the general economy. Distinguishing characteristics of agro-industries are embedded in the type and nature of risk inherent in the sector. In formulating public policies to mitigate against such uncertainties, thereby creating an enabling environment, we established a hierarchy of enabling needs for agro-industry competitiveness to inform public policy makers. We then went beyond this linear hierarchy to discuss the dynamic role of the state based on sector risk and capacity that must be considered in reforming public policy.

In recent decades, agrifood markets globally have experienced rapid change. This dynamic environment has raised the premium on traditional policy approaches to the sector, where 'getting agriculture moving' is now being replaced with 'making markets work' to sustain economic progress. Previously, ministers of agriculture were myopically focused on increasing on-farm productivity without paying closer attention to the enabling environment that addressed the competitiveness of the sector. While there is wide variation across countries and sectors, a 'one size fits all' strategy would be inappropriate to advance a reform agenda for the agrifood industries in developing countries. For effective reform to emerge, we therefore advocate a more nuanced appreciation of the role public policy makers can play in sustaining competitiveness.

Creative public policy in this dynamic global economy seeks to sustain efficient and equitable outcomes for the agrifood sector that call for government to develop essential enablers, promote innovative institutions, advance facilitative policies and maintain competitiveness. We recognize that, while agriculture is unique, it too must exist in a wider national economy. Therefore, our list of specific policy measures that are essential, important and useful to the agro-industry of developing counties must be coordinated with the wider national macroeconomic policy framework.

Establishing the 'rules of the game' in the form of property rights, especially in the case of deeds for physical and intellectual property, is a critical aspect of an enabling environment for agro-industries. Included as an essential enabler is contract enforcement. Given the rise of contract farming, vertical coordination and supply chain management of large food companies, efficiency and equity in the sector are undermined without strong laws to ensure the transaction implied within contracts in agriculture. The disadvantaged small farmer engaged in contract farming can become an efficient player in the market with strong laws that enforce contracts made with large agro-industries, and likewise companies that offer contracts can be assured of delivery of goods and services. Enforcement of contracts used by agro-industries is, however, part and parcel of 'the rule of law' established by any nation. Therefore, ministries of agriculture must aggressively and urgently expand such legal remedies to a wider set of transitions in rural areas.

Rural areas within emerging economies are often confronted with highly risky environments and weak public institutions. Under such circumstances, segments of the economy can exhibit market failure in the face of state failure. The private sector does not develop an appetite for investing in such environments, and the state lacks capacity to improve this environment such that inward investments could be made. The promotion of innovative institutions is critical to enhancing the bargaining power of farmers. The position of farmers in the market needs strengthening in most countries. Investment in farmers' associations can reduce transaction costs, enhance bargaining with suppliers and stimulate on-farm production.

References

Cecil, C.O. (2005). Gum arabica. *Suadi Aramco World 56*(2). Available at: http://www.saudiaramcoworld.com/issue/200502/gum.arabic.htm.

Crotty, A. (2007). Tiger Brands' price-fixing scandal a drop in the ocean. *Business Report*. Available at http://www.busrep.co.za/index.php?fArticleId=4146875.

Feder, G. & Feeny, D. (1991). Land tenure and property rights: theory and implications for development policy. *World Bank Economic Review 5*, 135–153.

Fennema, O.R. (1996). *Food Chemistry*. New York: Marcel Dekker.

Gutman, G.E. (2002). Impact of the rapid rise of supermarkets on dairy products systems in Argentina. *Development Policy Review 20*(4), 409–427.

Herdt, R.W. (2009). Overcoming poverty through improved agricultural technology. In R.D. Christy (ed.) *Financial Inclusion, Innovation and Investments*, ch. 4 (in press).

Hu, D., Reardon, T., Rozelle, S., Timmer, P. & Wang, H. (2004). The emergence of supermarkets with Chinese characteristics: challenges and opportunities for China's agricultural development. *Development Policy Review 22*(5), 557–586.

Knight, F.H. (1921). *Risk, Uncertainty and Profit*. New York: Houghton Mifflin.

Muhhuku, F. (2002). Seed industry development and seed legislation in Uganda. In N.P. Louwaars (ed.) *Seed Policy, Legislation, and Law: Widening a Narrow Focus*. Binghamton, NY: Food Products Press, an imprint of The Haworth Press, Inc., pp. 165–176.

National Agricultural Marketing Council (2003). *Food Pricing Monitoring Committee – Final Report*, Report prepared for the National Agricultural Marketing Council, Pretoria, South Africa.

Organization for Economic Cooperation and Development (OECD) (1992). *Technology and the Economy: the Key Relationships*. Paris.

Organization for Economic Cooperation and Development (2006). *Policy Framework for Investment*. Paris: OECD.

Pardey, P.G. & Beintema, N.M. (2001). *Slow Magic*. Food policy reports 13. International Food Policy Research Institute.

Peters, H.J. (1992). *Service: The New Focus in International Manufacturing and Trade*. World Bank Policy Research Working Paper 950. Washington, DC: Infrastructure and Urban Development.

Porter, M.E. (2004). Building the microeconomic foundations of prosperity: findings from the Business Competitiveness Index. In *The Global Competitiveness Report, 2004–2005*. New York: World Economic Forum, pp. 29–56.

Reardon, T., Timmer, P. & Berdegue, J. (2004). The rapid rise of supermarkets in developing countries: induced organizational, institutional, and technological change in agrifood systems. *The Electronic Journal of Agricultural and Development Economics, Food and Agriculture Organization of the United Nations 1*(2), 168–183.

Seria, N. (2007). Tiger Brands admits to bread price-fixing, pays fine. *MoneyWeb* November 13, 2007, Moneyweb Holdings Limited.

Ruttan, V.W. (1975). Technology transfer, institutional transfer, and induced technical and institutional change in agricultural development. In L.G. Reynolds (ed.) *Agriculture in Development Theory*. Yale University Press: New Haven, Connecticut, pp. 165–191.

World Bank (1995). *Investment Climate Surveys*. Washington, DC: World Bank.

World Bank (2003). *Doing Business in 2004: Understanding Regulation*. Washington, DC: World Bank and Oxford University Press.

World Bank (2007). *Doing Business 2008*. Washington, DC: World Bank and the International Finance Corporation.

World Economic Forum (2000). *The Global Competitiveness Report 2000*. New York: Oxford University Press.

World Economic Forum (2004). *The Global Competitiveness Report 2004–2005.* New York: Palgrave Macmillan.

World Economic Forum (2006). *The Global Competitiveness Report 2006–2007.* New York: Palgrave Macmillan.

World Economic Forum (2007). *The Global Competitiveness Report 2007–2008.* New York: Palgrave Macmillan.

Appendix 1. Ease of Doing Business indicators. (From World Bank, 2007.)

Business stage	Indicator	Measures
Starting a business	Procedures (number)	Number of procedures required to start a business
	Time (days)	Total amount of time required to complete all procedures
	Cost (% of income per capita)	Logistics costs as a percentage of annual income per capita
	Minimum capital (% of income per capita)	Paid-in minimum capital requirement as percentage of annual income per capita
Dealing with licences	Procedures (number)	All procedures required for a business in the construction industry to build a standardized warehouse
	Time (days)	Number of calendar days required to complete licensing procedures
	Cost (% of income per capita)	Official costs as a percentage of the country's annual income per capita
Employing workers	Difficulty of hiring index (0–100)	If fixed contracts are prohibited for permanent tasks, maximum cumulative duration of fixed-term contracts, ratio of minimum wage to average value added per worker (using a 42-year-old, non-executive, full-time, male employee as model worker)
	Rigidity of hours index (0–100)	Restrictions with regard to workweek hours, night work, weekend work and vacation days
	Difficulty of firing index (0–100)	Restrictions with regard to basis and procedure for terminating employment, laws pertaining to reassignment options and priority rules
	Rigidity of employment index (0–100)	Average of the three sub-indices above
	Non-wage labour cost (% of salary)	All social security payments and payroll taxes for the fiscal year
	Firing cost (weeks of salary)	Cost of advance notice requirements, severance payments and penalties due when terminating a redundant worker
Registering property	Procedure (number)	Number of procedures required to register or transfer property (standardized to land and building of 50 times income per capita value, under same ownership for 10 years)
	Time (days)	Total amount of time required for property registration
	Cost (% of property value)	Official costs associated with property registration as a percentage of property value
Getting credit	Strength of legal rights index (0–10)	Extent to which collateral and bankruptcy laws protect rights of borrowers and lenders, thus facilitate lending
	Depth of credit information index (0–6)	Rules affecting the scope, accessibility and quality of credit information available on borrowers

continued

Appendix 1. Continued.

	Public registry coverage (% of adults)	Number of individuals and firms listed in a public credit registry, current information on repayment history, unpaid debt and outstanding credit
	Private bureau coverage (% of adults)	Number of individuals and firms listed by private credit bureaus with current information on repayment history, unpaid debt and outstanding credit
Protecting investors	Extent of disclosure index (0–10)	Disclosure of related-party transactions (firm standardized to publicly traded corporate buyer, a food manufacturer)
	Extent of director liability index (0–10)	Clear obligations and codes of conduct for company directors and managers, and the shareholders' ability to successfully sue in case of fraud or bad faith
	Ease of shareholder suits index (0–10)	Easy access to the courts when investors are harmed (availability of documentation and witnesses from the firm during trial)
	Strength of investor protection index (0–10)	Average of the three sub-indices above
Paying taxes	Payments (number per year)	Total number of taxes and contributions paid (including corporate income tax, VAT, social contributions and taxes on labour, property and property transfer, dividends, capital gains, financial transactions, waste collection, vehicles, roads) for a standardized manufacturing company
	Time (h per year)	Time required to comply with three major taxes: income taxes, VAT and labour taxes
	Total tax rate (% of profit)	Total taxes as percentage of commercial profits
Trading across borders	Documents to export/import (number)	Trading documents to be filed (for a standardized 100+ employee firm exporting >10% of sales; product is traded in a dry cargo, 20 ft full container load)
	Time to export/import (days)	Time needed to comply with export and import requirements
	Cost of export/import (US$ per container)	Cross-border trade costs per container, including all fees associated with completing the procedures to trade (excludes tariffs and trade taxes)
Enforcing contracts	Procedures (number)	Number of litigation procedures necessary to resolve a commercial dispute
	Time (days)	Time required to resolve a contract dispute in court, i.e. from case filing to payment
	Cost (% of claim)	Cost of court and attorney fees as percentage of amount claimed
Closing a business	Time (years)	Time required to liquidate unviable businesses
	Cost (% of estate)	Cost of liquidation as percentage of estate value
	Recovery rate (cents on the dollar)	Proportion of loan recovered by creditors through the bankruptcy proceedings

Appendix 2. Doing business in Egypt, 2006/07. (From World Bank, 2007.)

	Components	In-stage rank	World rank
1. Starting a business	Procedures (7) Time (9 days) Cost (28.6% of income per capita) Minimum capital (12.9%)	55	126
2. Dealing with licences	Procedures (28) Time (249 days) Cost (474.9% of income per capita)	163	

3. Employing workers	Difficulty of hiring (0/100)
	Rigidity of hours (20/100)
	Difficulty of firing index (60/100)
	Rigidity of employment
	    index (27/100)
	Non-wage labour cost
	    (25% of salary)
	Firing cost (132 weeks of salary)
		108	

4. Registering property	Procedures (7) Time (193 days) Cost (1% of property value)	101	

5. Getting credit	Strength of legal rights (1/10)
	Depth of credit info index (4/6)
	Public registry coverage (1.6%
	    of adults)
	Private bureau coverage
	    (missing % of adults)
		115	

6. Protecting investors	Disclosure index (7/10) Director liability index (3/10) Ease of shareholder suits index (5/10) Investor protection index (5.0/10)	83	
7. Paying taxes	Procedures (36) Time (711 h per year) Cost (47.9% of profit)	150	
8. Trading across borders	Documents to export/import (6/7) Time to export/import (15/7) Cost of export/import ($714/$729 per container)	26	
9. Enforcing contracts	Procedures (42) Time (1010 days) Cost (25.3% of claim)	145	
10. Closing a business	Time (4.2 years) Cost (22% of estate) Recovery rate (16.6 cents on the dollar)	125	

Computation Procedure

Consider the ease of property registration in Egypt in the 2006/07 period: registration entailed completing seven different procedures, a process that took at least 193 days and would cost an equivalent of 1% of the value of the property to be registered. For each of these values, Egypt is ranked against other world economies and assigned a percentile ranking of 58.7%, 86.4% and 11.8%, respectively. Taking a simple average of these values gives the stage-level percentile ranking of 52.3% for Egypt's property registration processes. The same process is repeated for each of the business life stages, to produce 10 stage-level percentile rankings (these easily can be translated into a 1–178 ranking relative to other countries' performance in performing the same operations as indicated above for Egypt), whose simple average is the country-level percentile ranking. This percentile ranking is used to assign an Ease of Doing Business position for Egypt in the world economy, in this case, 126 out of 178.

Appendix 3. The Global Competitiveness Index (GCI). (From World Economic Forum, 2006.)

Competitiveness pillar	Measures	Components
Basic requirements		
Institutions	Rules that shape incentives and define the way economic agents interact in an economy	Public institutions: public-sector accountability, efficiency, transparency; independence of judiciary; respect of property rights; government inefficiency; levels of public security Private institutions: corporate ethics; accountability
Infrastructure	Quality of physical infrastructure, i.e. energy, transport and telecommunications	Overall infrastructure quality; railway infrastructure development; quality of port infrastructure; quality of air transport infrastructure; quality of electricity supply; telephone lines
Macroeconomy	Level of macroeconomic stability	Fiscal indicators (government deficit, national savings rate, inflation, interest rates, government debt); trade-weighted real effective exchange rate
Efficiency enhancers		
Health and education	Quality of primary health and education	Health: medium-term business impact of malaria, tuberculosis and HIV/AIDS; infant mortality; life expectancy; TB prevalence; malaria prevalence; HIV prevalence Education: primary school enrolment

continued

Appendix 3. Continued.

Competitiveness pillar	Measures	Components
Higher education	Quality of educational system	Secondary and tertiary enrolment rates; quality of education as assessed by business community (quality of science, maths and management schools); availability of specialized training for workforce
Market efficiency	Extent to which goods, labour and finance are allocated efficiently for maximum productivity	Goods: market openness, level of distortive government intervention; market size Labour: cooperation in employer–employee relations; flexibility of labour regulations; extent of gender bias in the workplace Financial: access to credit; quality of capital; soundness of banking sector
Technological readiness	Agility with which an economy adopts existing technologies to enhance the productivity of its industries	Availability of information and communications technologies (ICTs) and other technologies; aggressiveness of firm adoption of new technologies; FDI and technology transfer; cellular telephones; Internet users; personal computers
Innovation and sophistication factors		
Business sophistication	Ability of business leaders to manage companies efficiently	Quantity and quality of local suppliers, level of development of production processes, extent to which companies are turning out the most sophisticated products, networks and supporting industries
Innovation	Extent of design and development of cutting-edge products and processes	Business investment in research and development; quality of scientific research; extent of collaboration in research between universities and industries; protection of intellectual property

Appendix 4. Components of the Business Competitiveness Index (BCI). (From Porter, 2004.)

I. Company operations and strategy
    Production process sophistication
    Nature of competitive advantage
    Extent of staff training
    Willingness to delegate authority
    Capacity of innovation
    Company spending on research and development
    Value chain presence
    Breadth of international markets
    Degree of customer orientation
    Control of international distribution
    Extent of branding
    Reliance on professional management
    Extent of incentive compensation
    Extent of regional sales
    Prevalence of foreign technology licensing

II. National business environment
    (a) Factor conditions
    Physical infrastructure
    Administrative infrastructure
    Human resources
    Technology infrastructure
    Capital markets

    (b) Demand conditions
    Buyer sophistication
    Sophistication of local buyers, products and processes
    Government procurement of advanced technology products
    Presence of demanding regulatory standards
    Laws relating to information and communications technologies (ICTs)
    Stringency of environmental regulations

    (c) Related and supporting industries
    Local supplier quality
    State of cluster development
    Local availability of process machinery
    Local availability of specialized research and training services
    Extent of collaboration among clusters
    Local supplier quantity
    Local availability of components and parts

    (d) Context for firm strategy and rivalry
    Incentives: extent of distortive government subsidies
    Favouritism in decisions of government officials
    Cooperation in labour–employer relations
    Efficacy of corporate boards
    Intellectual property protection
    Protection of minority shareholder interests
    Regulation of securities exchanges
    Effectiveness of bankruptcy laws

continued

Appendix 4. Continued.

Competition: hidden trade barriers
Intensity of local competition
Extent of locally based competitors
Effectiveness of antitrust policy
Decentralization of corporate activity
Business costs of corruption
Costs of importing foreign equipment
Centralization of economic policy making
Prevalence of mergers and acquisitions
Foreign ownership restrictions

Appendix 5. Components of the Investment Compass (IC). (From UNCTAD web site http://compass.unctad.org/Page1.egml?country1=&country2=&region=&sessioncontext=246934608&object=SC.app.objects.methodology.)

Key area	Variables	Indicators
Resource assets	Human capital	School enrolment in tertiary
		Science and engineering students
		Illiteracy rate
		Production of minerals
	Availability of raw materials	Production of agricultural commodities
		Energy reserves
	Market size	GDP (purchasing power parity) (PPP)
		Per capita income (PPP)
		Effective market size
		Total population
Infrastructure	ICT	Internet hosts
		Internet users
		Mobile phones
		Telephone main lines
	Basic infrastructure	Air transport freight
		Water access
		Electricity transmission and distribution losses
		Electricity production
		Port activity
		Road network
		Railway freight
Operating costs	Labour costs	Monthly wage for professional work
		Monthly wage for administrative work
		Monthly wage for technical work
		Monthly wage for clerical work

continued

Appendix 5. Continued

Key area	Variables	Indicators
	Business costs	Rental office costs
		International telecommunications charge
		Local telecommunications charge
		Electricity charge
Economic performance and governance	Macroeconomic performance	Unemployment rate
		Government surplus/deficit
		Current account balance
		Inflation
		Real economic growth
	Governance	Creditworthiness rating
		Human development index
		Regulatory quality
		Rule of law
		Government effectiveness
		Political stability
		Voice and accountability
Taxation and incentives	Business and professional services (present value of tax)	Business and professional services (direct and indirect)
	Information and communications technology (present value of tax) (PV tax)	ICT (direct and indirect)
	Tourism (PV tax)	Tourism (direct and indirect)
	Manufacturing (PV tax)	Manufacturing (direct and indirect)
Regulatory framework	Entry	Standards of treatment
		Land ownership and transfer
		Openness of main sectors to FDI
	Operation	Performance requirements
		Preferential trade arrangements with major markets
		Labour market regulation
		Foreign workforce regulations
		Foreign exchange regulation on current operations
		Size of regional/integrated trade area
		Import duties
	Protection and exit	Number of taxation treaties signed
		Number of bilateral investment treaties signed
		Liquidation and expropriation
		Dispute settlement

Appendix 6. Components of the Policy Framework for Investment. (From Organization for Economic Cooperation and Development, 2006.)

Policy field	Rationale	Major issues
Investment policy	Quality of investment policies directly influences investment decision, driven by principles including transparency, property protection and non-discrimination	Enforcement, transparency, accessibility of laws and regulations Efficient ownership registration of property Protection of intellectual property Effective contract enforcement Effective compensation expropriation Non-discrimination principle Implementation of international arbitration instruments
Investment promotion and facilitation	Provide aim to correct for market failures and leverage the strong point of a country's investment environment	Government strategy for sound business environment Establishing and adequately funding investment promotion agency Streamlining administrative procedures for new investments Maintaining dialogue with investors Evaluating costs and benefits of investment incentives Facilitating investment linkages between businesses
Trade policy	Support more and better-quality investment by expanding opportunities to reap scale economies and by facilitating integration into global supply chains	Reducing border costs and inefficiencies Reducing trade policy uncertainty Participation in international trade agreements Review of trade policy to reduce distortions Impact of trade policy on input prices Alternative means of achieving public policy objectives Targeting policy to attract investment towards weak sectors
Competition policy	Favours innovation and contributes to conditions conducive to new investment	Clear, transparent, non-discrimination competition policy Are competition authorities adequately resourced? Ability to address anticompetitive practices Capacity to evaluate impact on market entry of other policies Capacity to evaluate the costs and benefits of industrial policies Role of competition authorities in case of privatization Extent of cooperation in international competition issues

continued

Appendix 6. Continued.

Policy field	Rationale	Major issues
Tax policy	Level of tax burden and design and administration of tax policy influence business costs and returns on investment	Average tax burden on domestic profits, accounting for statutory provisions, tax-planning opportunities and compliance costs Tax burden consistent with investment attraction strategy Tax burden consistent with goals and objectives of tax system Neutrality of tax system to nationality of investor, firm size, age of business entity, ownership structure, industry sector, location Consistency of main tax provisions with international norms Presence of unintended tax opportunities resulting from targeted tax incentives, and their impact on cost-effectiveness of system Tax expenditure account reporting and use of sunset clauses to inform and manage the budget process Extent of tax treaty network and presence of strategies to counter abusive cross-border tax-planning strategies
Corporate governance	Influences confidence in investors, cost of capital, overall functioning of financial markets and development of more sustainable sources of financing	Presence of coherent, consistent regulatory framework backed with effective enforcement Extent to which framework ensures equitable treatment of shareholders Institutional structure for legal redress in case of violation of shareholder rights Procedures and institutions for shareholder empowerment Standards and procedures for timely, reliable and relevant disclosure Does framework ensure effective monitoring of management by the board? Voluntary incentives and training to encourage and develop a good corporate governance culture Has national corporate governance system been reviewed? For state-owned enterprises, extent of government interference in management and market operations
Policies for promoting responsible business conduct	Good conduct policies (respecting human rights, environmental protection, labour relations and financial accountability) help attract enterprises that contribute to sustainable development	Extent to which responsibilities ascribed to the business sector are clear Steps taken to communicate responsible business behaviour to investors Presence of framework to support company disclosures about business operations

continued

Human resource development	Policies that develop and maintain a skilled, adaptable and healthy population and ensure the full and productive deployment of human resources, thus a favourable investment environment	Government support of company efforts to comply with the law Government role in strengthening the base case for responsible conduct Government participation in intergovernmental cooperation to promote international principles for responsible business Presence of coherent and comprehensive human resource development policy framework Strategies for increased participation in basic schooling and to improve the quality of instruction so as to leverage assets Incentives for individuals to invest in higher education and lifelong learning Extent to which government promotes training programmes and evaluates effectiveness of investment environment Presence of coherent strategy to tackle the spread of pandemic diseases Mechanisms to promote and enforce core labour standards Extent to which labour market regulations support job creation and the government's investment attraction strategy Steps taken to unwind unduly restrictive practices covering the deployment of workers from investing enterprises Programmes to assist large-scale labour adjustment Steps to ensure that labour regulations support an adaptable workforce
Infrastructure and financial sector development	Ensure scarce resources are channelled to the most promising projects and address bottlenecks that limit private investment	Processes used to evaluate infrastructure investment needs Measures adopted to uphold transparency and procedural fairness in bidding for infrastructure development contracts Market access for potential investors in telecommunications and extent of competition Access to electricity services on a least-cost basis for a wide range of users Processes for development and maintenance of transport infrastructure Investment needs and private-sector involvement in water management Capacity of financial sector and quality of regulatory framework Laws and regulations governing credit access

Appendix 6. Continued.

Policy field	Rationale	Major issues
Public governance	Regulatory quality and public-sector integrity important for establishing credibility with investors and for reaping development benefits of investments	Presence of a coherent and comprehensive regulatory reform framework
		Mechanisms for managing and coordinating regulatory reforms across different levels of government
		Extent to which regulatory impact assessments are used to evaluate the consequences of economic regulations for the investment environment
		Public consultation mechanisms and procedures established to improve regulatory quality
		Extent to which the administrative burdens on investors are measured and qualified
		Extent to which international anticorruption and integrity standards have been implemented in national legislation
		Extent to which institutions and procedures ensure transparent, effective and consistent enforcement of anticorruption laws and regulations
		Existence of review mechanisms to assess the performance of laws and regulations on anticorruption and integrity
		Government participation in international initiatives aimed at fighting corruption and improving public-sector integrity

Appendix 7. Components of the core questionnaire for Investment Climate Surveys. (From World Bank web site.)

Section	Issues
1. General information	Age of enterprise, legal status, ownership, number of operating facilities, main product line, other income-generating activities
2. Sales and supplies	Share of local market, percentage of sale into different markets, source of inputs, days of inventory at hand, supply delivery delays, competitors in domestic market
3. Investment climate constraints on the firm	Severity of obstacles in telecommunications, electricity, transport, access to land, tax rates, tax administration, customs and trade regulations, labour regulations, skills and education of available workers, business licensing and operating permits, access to financing, cost of financing, economic and regulatory policy uncertainty, macroeconomic instability, corruption, crime, anticompetitive or informal practices, conflict resolution
4. Infrastructure and services	Frequency and duration of service interruptions, average utility costs, access to computer and the Internet, affiliation to chamber of commerce and services received, quality and affordability of business services
5. Finance	Contribution to working capital and new investments of finances from different sources (e.g. internal funds, loans, investment funds, rental income), access to credit or overdraft facilities, interest rate on loans, frequency of external auditor reviews
6. Business–government relations	Efficiency of government in delivering services, consistency and predictability of officials' interpretation of regulations, days to import/export, restrictiveness of labour regulations, time spent complying with regulations, bribes as percentage of annual sales, percentage of sales reported for taxes, influence on national laws and regulations of lobbying efforts by various entities (firms, business associations, labour unions, local government)
7. Conflict resolution and legal environment	Confidence in judicial system, proportion of sales on credit, percentage of sale to government and state-owned enterprises, time spent in payment dispute resolution, proportion of disputes resolved by court action
8. Crime	Percentage of sale spent on security and protection payments, losses due to theft, vandalism or arson, share of crimes reported to the police and share solved
9. Capacity, innovation, learning	Average capacity utilization, rate of growth in sales, share of profits reinvested, range of products, use of technology from foreign-owned firms, position on technology relative to competitors, innovation initiatives recently taken by firm, ways of acquiring technological innovations, sources of pressure to reduce production costs and to develop new products and services
10. Labour relations	Average number of workers, education levels, total compensation, proportion of foreign employees, time spent recruiting, formal training offered to employees, percentage of unionized workforce, frequency of labour disputes, proportion of female employees, level of top management experience
11. Productivity	Total sales; input costs; market value of production; energy costs; manpower costs; interest charges and financial fees; spending on new assets; value of assets sold; amount spent on rentals, leases and licences; amount spent on research and development; net value of assets; value of liabilities

6 Business Models That Are Inclusive of Small Farmers[*]

Bill Vorley,[1] Mark Lundy[2] and James MacGregor[3]

[1]Head, Sustainable Markets Group, International Institute for Environment and Development (IIED), London, UK; [2]Agroenterprise Specialist, International Center for Tropical Agriculture, CIAT, Cali, Colombia; [3]Researcher, International Institute for Environment and Development (IIED), London, UK

Introduction

Small-scale farmers, who form the bedrock for global agrifood supply, are faced with markets in an unprecedented state of flux. Domestic markets are undergoing rapid but uneven modernization, and higher value and export markets are increasingly the preserve of larger-scale suppliers.

The modernization of domestic markets, particularly in Latin America and Asia, has been driven by a wave of investments in emerging economies by domestic and transnational food manufacturers and retailers over the past two decades. Combined with rising urbanization, and changes in consumer preferences and purchasing power, these have led to a growth of modern organized food retailing, which has outpaced the growth of per-capita GDP by a factor of 3–5 (Reardon and Huang, 2008).

These changes are generating intense policy debate, particularly regarding the opportunities facing small farmers and the rural poor. The 2008 World Bank World Development Report (WDR) (World Bank, 2007b) notes that in transforming economies, where the majority of the rural poor live, 'the rising urban–rural income gap accompanied by unfulfilled expectations creates political tensions. Growth in agriculture and the rural non-farm economy is needed

*This chapter draws heavily on the work of the Regoverning Markets consortium (www. regoverningmarkets.org) and the associated international conference 'Inclusive Business in Agrifood Markets: Evidence and Action' held in Beijing on 5–6 March 2008. Contributions are also acknowledged from the 'New Business Models for Sustainable Trade' project led by the Sustainable Food Laboratory and Rainforest Alliance, 'Inclusion of small producers in value chains' – a partnership between Cordaid, Vredeseilanden and IIED; and framework funding to IIED from the Swedish International Development Agency (SIDA). Comments on an earlier draft by Jose Reijter (Cordaid) are gratefully acknowledged.

to reduce rural poverty and narrow the urban–rural divide.' Those political tensions are clear in India, where the fragmented US$350 billion retail industry is forecast to double in size by 2015, and where modernization and liberalization of retail foreign direct investment (FDI) have given rise to heightened investment coupled with significant protest and policy push-back.

Market modernization can offer increased economic opportunities for producers, consumers, entrepreneurs and other actors in the food chain. These opportunities include a reduction in entry barriers to traditionally protected industries, which are further leveraged by clearer information, less capture by elites, stronger access to services and the potential for entrepreneurial farmers to combine resources and realize the collective worth of their land. In some areas, new market entrants are stimulating competition for farmers' produce, helping to increase the value retained in rural economies. For example, the laws which entrenched a monopoly of wholesale markets in India have been amended in at least 14 states, allowing retailers and their agents to procure directly from farmers.[1] Enforced intermediation through wholesale commission agents had previously hidden the final buyer from farmers.

But there are also risks in opening up markets, where domestic businesses may be by-passed by cheaper imports and where costly market entry requirements favour the better-resourced. These features, which have long been understood in export markets, are becoming a feature of domestic markets in emerging economies as regional trade becomes easier.

If the benefits of modernization and globalization are patchy, and do not reach to the 'bottom of the pyramid' to deliver a growth and equity 'win–win', then prospects for meeting the Millennium Development Goals (MDGs) by 2015 are remote. The 2008 WDR calls for action in response to the modernization of procurement systems in integrated supply chains and supermarkets, so that small-scale farmers can share in these growth opportunities.

The failure of major retailers to take such a combined 'growth with equity' approach was bemoaned by the late Robert Davies, former CEO of the International Business Leaders Forum (IBLF). He asked: 'Why are the cleverest logistics and supply chain operators and service companies known in business history sometimes so inept at... adapting their business model to the sensitivities of emerging markets?' (Davies, 2007) Part of the answer – and the subject of this chapter – lies in the development of business models which are both inclusive of small-scale producers and also address the need for processors and retailers to manage costs and risks.

Here we define inclusive business models as those which do not leave behind small-scale farmers and in which the voices and needs of those actors in rural areas in developing countries are recognized. Such models have been variously described as 'Inclusive Business' (WBCSD and SNV, 2008; www. inclusivebusiness.org), 'Mutually Beneficial Partnerships' (FAO and CIFOR, 2002) and *'inclusive capitalism'* (Hart, 2007).

---

[1] See 'Modern retail offers wide choice, farmers want to exercise it all'. Livemint.com/*Wall Street Journal* 11 February 2008.

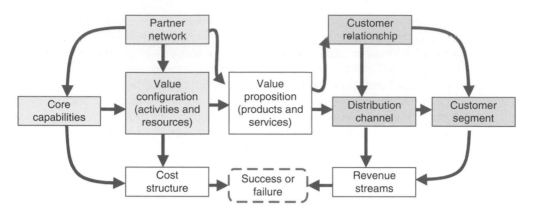

Figure 1. Template of a business model. (Adapted from Osterwalder, 2006.)

The chapter describes a range of business models for inclusive market development within the context of agrifood restructuring and modernization. It focuses specifically on models that improve the inclusiveness, fairness, durability and financial sustainability of trading relationships between small farmers on the one hand and downstream agribusiness (processors, exporters and retailers) on the other. It also alerts us to the needs of external providers, such as financiers and training agents. The gap in basic services in rural economies, such as appropriate extension and credit, needs to be bridged before FDI can live up to its promises. While we do address what producers need to do to compete in modern dynamic markets, and the role of facilitating public policy, our focus in this chapter is more on the buyers and their role as partners in development.

Business Models and Inclusive Market Development

What is a business model?

A *business model* is the way by which a business creates and captures value within a market network of producers, suppliers and consumers, or, in short, 'what a company does and how it makes money from doing it' (MIT Sloan[2]).

The business model concept is linked to business *strategy* (the process of business model design) and business *operations* (the implementation of a company's business model into organizational structures and systems). Osterwalder (2006) breaks business models into their constituent elements that create costs and value, using the template in Figure 1.

This template shows the importance of *market differentiation* (building a 'value proposition') and *cost management* to the success of any business model.

---

[2] http://process.mit.edu/Info/eModels.asp

In modern agrifood retail, market differentiation is built on consumer assurance, high standards for food quality and safety, year-round availability and, sometimes, lower prices that are communicated to consumers through own brands.

It follows that the *partner network* – the supply chain and its coordination – is a vitally important source of competitive advantage. It also follows that the model is highly sensitive to any addition of costs and risks, and it is around this apex that the question of market inclusivity ultimately revolves.

There are perceived to be high transaction costs and increased risks with purchasing from large numbers of fragmented small-scale farmers. Small-scale farmers are also perceived to be less reliable in honouring trading agreements, because they do not have the technical skills and technologies to produce the right products at the right time (quality, timeliness and consistency). Common business practice sees buyers typically seeking out large-scale suppliers (Box 1) and areas favoured by agribusiness, such as zones involved in export production. This is particularly easy in a dualistic farm structure such as that found in South Africa (Box 2).

From the perspective of producers and their organizations, there may be good reasons to avoid trading with the modern agrifood system. With low and inconsistent production volumes, dispersed production, weak negotiation positions, limited capacity to upgrade and meet formal market

---

Box 1. Carrefour's quality line in China. (From Hu and Xia, 2007.)

Among the supermarkets in China, Carrefour is characterized by marketing fresh foods. With the rising consciousness of consumers on safety of food, the demand for high-quality and safe food has increased. Carrefour started in 1999 to sell a 'green' food supply line under its own brand with the 'Quality Food Carrefour' logo. These lines represent an innovation in the purchasing system within the Chinese context, where Carrefour carries out integrated management of the entire supply chain, with full traceability. Other retailers are following suit. To date, cooperators of the Carrefour quality line are all larger-scale, rather than the small-scale farmers who account for more than 90% of the agricultural population in China.

Box 2. From wholesale to preferred supplier: Shoprite.

Shoprite, a leading South African retailer, relied on sourcing from wholesale markets in 1992 for 70% of its produce. In 1992, Freshmark, a wholly owned specialized and dedicated wholesaler, started to form 'preferred supplier' relationships with large commercial farmers (from whom it sources the majority of its produce), as well as some large wholesalers and some medium- and smaller-scale farmers. By 2006, it had 700 such preferred suppliers (a few for each main product), and sourced 90% of its produce from them and 10% from the wholesale markets. The shift to using preferred suppliers was facilitated in South Africa by the sharply dualistic farm sector structure. Freshmark has 'followed' Shoprite into other African countries, but is still sourcing much of its produce from South Africa.

requirements and poor access to information, technology and finance, the transaction costs for farmers to link with the modern sector are daunting. And despite significant investments of time and resources, market access is still not guaranteed.

Ultimately, the type of partner network and choice of business model will depend on the nature of the product (perishable, differentiated or branded product or bulk commodity) and the nature of the end buyer (branded retailer, wholesaler, etc.), which determine the nature of economic dependency between chain actors. A *collaborative partner network* is much more important with perishable commodities such as fresh vegetables, dairy and meat, which require traceability and have higher food safety risk profiles (Sporleder *et al.*, 2005). The same applies to the growing number of certified products, such as Fairtrade and organics. Jan Van Roekel of the Agro-Chain Competence Foundation goes as far as to say: 'In the future, agrifood producers, processors and retailers will no longer compete as individual entities. Rather, they will collaborate as a strategic value chain and compete with other value chains in the market place' (Bouma, 2005).

Crucially for the discussion of inclusion of small-scale farmers, these collaborative partner networks, with co-investment and knowledge sharing between producers, suppliers, processors and retailers, are usually built around a small number of preferred suppliers. Adapted business models are called for, whereby small-scale farmers can cooperate to compete as one single supplier, and where their customers are responsive to the realities of smallholder production.

Adapting business models

When agrifood business models are transplanted from industrialized countries to countries with large agriculture-dependent populations the unintended consequences for the rural economy can potentially be very significant.

In the two largest 'transforming' countries,[3] China and India, 40–60% of the workforce is engaged in agriculture, i.e. over 640 million people in total. In Thailand, Turkey and Morocco this figure is 40–50% of the workforce, while in Romania and Honduras agriculture still accounts for one-third of employment. The small-scale retail sector also faces major challenges. The Indian retail sector, dominated by 15 million very small independent *kirana* stores, employs 42 million people, the second biggest employer after agriculture. In addition to the lack of inclusiveness and poor awareness of rural economic realities of existing business models, there are other unintended and accidental outcomes. For example, the private standards that buyers have deemed necessary and efficient solutions can compound exclusion of small suppliers given that the costs of such systems are a function of production volume. Put simply, high volume makes standards feasible because the cost per unit of product is low. For many small-

---

[3] One of agriculture's three worlds, according to WDR 2008, in which agriculture contributes less to growth, but poverty remains overwhelmingly rural.

holders, however, volumes are low and the unit cost for standards is high. This mix can lead to situations where it is financially impossible for smallholders to cover the cost of implementing and maintaining such standards.

It is clear that new business models are needed that afford opportunity in terms of small-farmer inclusion and equity and that do not exclude efficient farmers, while promoting business efficiencies. There are potential efficiency gains in developing locally adapted business models that build on the comparative advantage of smallholders, in terms of land, price, farm management, quality and innovation. According to the template of business models above, any adjustments in pursuit of greater inclusiveness must not undermine the most sensitive elements of a model – the cost structure, the value proposition and the integrity and safety of the product – especially when managing supply from large numbers of small producers.

What Is the Business Case for Adjusting Business Models in Favour of Smallholders?

The first section has described how the business model of the organized agrifood sector is generally built on the value proposition of consumer assurance, high standards for food quality and safety, low prices and reliability of supply. It sets out the biggest challenge for modern agrifood business to work with small-scale farmers as being to organize supply to deliver the benefits of logistics, economies of scale, traceability and private sector standards.

While the business case for trading with small-scale producers is being called into question, experience in the field suggests that a convincing business case for models that are inclusive of small-scale producers can be made beyond efforts to promote corporate social responsibility (CSR), based primarily on securing supply and reducing costs. Table 1 provides a summary of the arguments for and against sourcing strategies built with small-scale producers.

Securing supply

Securing consistent supply is especially critical in supply-constrained and volatile conditions, such as those currently characterizing global agrifood markets. A shift from a buyer's to a seller's market implies that suppliers will need to ensure that they can meet their obligations to retailer or processor customers in the face of considerable uncertainty. A diversified supplier base, including small-scale producers, can contribute to improved security of supply.

Retail buyers and processors may also seek to by-pass markets where large traders have a stranglehold. This was the situation in Pakistan where a milk processor, Haleeb Foods Limited, worked around the large and well-established milk traders by securing a small-farmer supply base.

Table 1. The business case for and against procuring from small-scale producers.

For	Against
• Smallholders' comparative advantages (premium quality, access to land, etc.) • securing supply in volatile markets, spreading portfolio geographically, reducing risk of undersupply as well as localized pest and disease problems • new business, clients for other products and services (base of pyramid) • new technologies available (efficient low-scale processing equipment, information technologies for coordination and lower cost traceability) • capacity to ramp up or ramp down production without incurring fixed costs (contract farming) • access to donor assistance • corporate responsibility • community goodwill • political capital	Costs and risks in organizing supply from dispersed producers: • quantity • quality • consistency • safety • traceability • compliance with rising standards • packaging • loyalty and fulfilment of commitments by farmers • negotiation time and costs • political opposition to commercialization of peasant agriculture

An even stronger business case for linking with small-scale producers is where there is a scarcity of alternative suppliers, whether due to the characteristics of the product (seasonality, labour requirements, locality), a shortage of land for large-scale domestic or own-business production, a lack of a medium- or large-scale supply base (e.g. the dairy sector in India or Poland) or where there is demand in remote areas away from main distribution channels (Box 4).

Small-scale producers can also have a comparative advantage in terms of produce quality, innovation, costs and farm management. Indeed, in exports of fresh vegetables from Africa to the UK and from Central America to the USA, it is the premium quality products such as French beans and peas that are sourced from smallholders.

New business opportunities

Small-scale producers are themselves a new business opportunity. In India retailers can now buy directly from farmers rather than operating through the government-controlled Agricultural Product Marketing Committee (APMC) wholesale markets. New models of rural retail are emerging, such as the Hariyali Kisaan Bazaar, which combines a 'bottom of the pyramid' approach to both the input and output sides of the farm-to-consumer value chain, and is discussed further in this chapter. This is an extension of the approach advocated by Prahalad and Hart (2002), which argues that corporations can make considerable profits by designing new business models and products to target the four billion poorest people who make up the base of the economic pyramid.

Box 3. Securing supply in remote regions.

Tanzania: Given the remoteness of tourist hotels, local supply from small-scale farmers is much less costly, especially during the rainy season when road transportation from outside the area is not always possible. Furthermore, the local supply has a promotional value in the tourist trade as a support to local communities, coupled with the encouragement of environmentally sound production (Mafuru *et al.*, 2007).

   South Africa: In contrast to the centralized fresh produce procurement systems of South African retailers who rely on preferred commercial suppliers, there are also innovative procurement schemes. Two rural-based supermarket chain stores in the Limpopo Province source fresh vegetables locally from small-scale farmers. By 2004, the Thohoyandou SPAR store was procuring approximately 30% of its vegetables from about 27 small-scale farmers. These farmers are supported by interest-free loans to selected farmers, a guaranteed market, farm visits and training on required quality standards. The remoteness of the supermarkets from the central distribution centres, the stores' operation in rural areas, reduced transportation costs and meeting freshness requirements, as well as being seen to contribute to community development, were the drivers for supporting the development of this local procurement scheme from small-scale farmers (Bienabe and Vermeulen, 2007).

Community goodwill

Working with small-scale farmers is also a means to build community goodwill, contributing to a company's licence to operate. Buying locally from smallholders may be part of a company's socially responsible strategy and becomes an advertising slogan in the highly competitive environment in which it operates. Customers are aware of and may value local procurement from small-scale farmers in the community as long as the produce is of a good quality. The case study from South Africa (Box 3) reported that the retailer organized for farmers to be present in the store on certain Fridays to promote its small-scale farmer procurement among the consumers.

   Other examples of supermarkets working with small-scale producers include programmes by Carrefour in Indonesia, through which dialogue is established between farmers and buyers to ensure increased quality, the development of a 'Best Supplier' prize and the waiving of listing fees.[4] In Guatemala, Wal-Mart has recently initiated a programme with an international NGO, Mercy Corps, and the financial service provider AGIL Foundation (Fundación Apoyo a la Generación de Ingresos Locales) to facilitate the entry of 600 farmers to its supply base over the next 3 years. The goal of this programme is to guarantee supplies of speciality products as Wal-Mart expands in the region. Finally, in Mozambique Shoprite, in collaboration with the IFAD Markets Support

---

[4] 'Carrefour Indonesia takes part in SME programme.' www.planetretail.net 1 August 2007.

Programme (PAMA), has supported the development of small-scale farmers in Boane. Based on this work, Shoprite now sources 25% of its fresh fruit and vegetable needs locally rather than importing from South Africa.

Despite the increasing interest of buyers to work with small farmers, questions remain about the depth of this commitment due to the fragmentary nature of some of these programmes. In Guatemala, for example, Wal-Mart executives have been important allies in the development of a line of personal care products based on medicinal plant extracts produced by indigenous communities in the municipality of Totonicapán. Despite this support, buyers and store-level display managers continue to obstruct the entry of these products in specific stores. The waiving of formal product registry by Wal-Mart executives has been used as an argument by lower-level staff for not including the product in display plans for specific stores, thus effectively keeping the products off the shelves despite high-level support (Lundy and Fujisaka, 2008). This example highlights the tensions and inconsistencies between executive desires and day-to-day business practices. Firms interested in promoting inclusive business models need to pay specific attention to the consistency of both their messages and their practice.

Corporate social responsibility

As discussed in Chapter 7 (this volume), which focuses on the theme of 'corporate social responsibility', an increasing number of companies report on their commitments to the development agenda to their customers and shareholders within a wider 'corporate responsibility' framework. The role of business as a partner in development has been a growing element of the CSR agenda, especially since the World Summit on Sustainable Development in 2002. It is now promulgated by a number of business platforms such as the World Business Council for Sustainable Development (WBCSD) and the Sustainable Agriculture Initiative (SAI), and by a number of UN agencies including UNDP and UNIDO. We are now at a point where 'inclusive business' and CSR concepts are being differentiated. CSR, with its emphasis on labour and environmental standards and supplier codes, has been poor at addressing market inclusion, and is often weakly mainstreamed across business. The UN Global Compact, which is the largest global corporate citizenship initiative, has ten principles that address human rights, labour standards, environment and anti-corruption, but do not address the role of business in supporting the position of primary producers. Some individual businesses and industries have gone further.

A commitment to the development agenda can defend a market. This has been evident recently in the UK, where airfreight of fresh produce from Africa was defended against a strong environmental critique through a clear demonstration of the importance of the trade to rural livelihoods (Garside *et al.*, 2007). However, it remains unclear how significant the commitment is, and whether these early actions will be followed up by the buyers, consumers or governments.

What Are the Various Models That Have Emerged for Linking Small-scale Farmers to Agribusiness and Changing Markets?

The preceding section elaborates the business case for inclusion. However, this will be insufficient to trigger widespread adoption of inclusive business models unless the risks and costs are addressed. Key to overcoming the costs and risks are producer coordination, market coordination and intermediation, service and finance provision, information and knowledge management and buyer behaviour. In this section we address producer coordination and market coordination of small-scale producers.

Organization of production is central to overcoming the costs associated with dispersion of producers, diseconomies of scale, poor access to information, technology and finance, inconsistent volume and quality, lack of traceability and management of risk. In view of lower transaction costs and the possibility of more effective capacity transfer, private companies often prefer to work with organized farmers rather than individuals despite the increased bargaining power that groups can enjoy. Production may be organized by the producers themselves, by the end customer companies or by an intermediary such as an NGO, trader, wholesaler or exporter (Table 2) in a range of direct or indirect market linkages categorized by Shepherd (2007). Organization might be layered, with buyers operating a continuum from preferred suppliers to top-up suppliers, with an attendant spread of objectives. Typically, success depends on communication flows and constant innovation on both sides.

Model development

Where market linkages are initiated by existing actors, they tend to build on informal structures in which traders or farmer-traders play a critical role not only to connect farmers to markets but also as *de facto* service providers. In

Table 2. Typical organization of smallholder production.

Type	Driver	Objective
Producer-driven	Small-scale producers themselves	• New markets • higher market price • stabilize market position
Buyer-driven	Large farmers Processors Exporters Retailers	• Extra supply volumes • Assure supply
Intermediary-driven	Traders, wholesalers and other traditional market actors	• Supply more discerning customers
	NGOs and other support agencies	• 'Make markets work for the poor'
	National and local governments, e.g. via 'Dragon Head' companies in China	• Regional development

many cases the trader is a member of the rural community and has specialized knowledge, information, assets and contacts to facilitate not only commercial ties but also social support in times of crisis. Informal linkage models are common throughout the world but little understood. Certainly, knowledge on how to develop business models that leverage these informal linkage systems is scarce. These models rarely receive support from development interventions or attention from researchers due to their informal nature and a strong bias against traders in many development organizations. This is unfortunate, as these models hold important information and lessons for sustainable market linkages and service provision, especially in areas with weak formal farmer organization.

Work in Colombia by one of the authors showed that traders are capable of extending market linkage services to smallholder farmers in a sustainable fashion when credit is provided (CIAT and CIPASLA, 2006). In some cases, aspects of trader-driven approaches are adapted by the private sector in lead-farmer models, such as those detailed in Honduras (Agropyme, 2006). A key finding is that informal market linkages are a form of cooperation or quasi-cooperation among farmers. For instance, informal moneylenders often hold extensive information about the needs, weaknesses and strengths of their customers, which moneylenders can leverage through supplying or by informing suppliers. Importantly, these forms of linkage/quasi-cooperation are the building blocks for formalized cooperatives.

A more traditional approach is small-farmer organization *induced* by external agents or a combination of external actors and small farmers. Processes of induced organizations start from the assumption that existing market linkages are not effective either in terms of efficiency or in terms of equity and that new skills and knowledge need to be developed to facilitate favourable market linkages for smallholders. These interventions are often led by development organizations and supported by donors although examples of private sector initiatives of induced organization, such as contract farming and outgrower schemes, also exist. Recent work raises doubts as to the sustainability of these induced organizations supported by development actors due to pressures to avoid failure (Berdegué, 2001), non-sustainable business practices (Hellin *et al.*, 2007; Shepherd, 2007) and inherent inefficiencies in the intervention model (Berdegué *et al.*, 2008a).

Despite the possibly poor performance of induced farmer business organizations led by NGOs and the public sector, there are cases where such interventions are effective especially where the facilitating organization has a strong business development focus. Of critical importance is a clear and consistent focus on the business case for the intervention as well as a timeline after which external support will cease. An example of this is the work carried out by the Presidential Commission for Local Development in Guatemala, which focuses on building 'business ecosystems' to support specific market opportunities. The Commission identifies and links key service providers to the supply chain as for-profit businesses rather than with donor subsidies. The resulting products and services incorporate support costs as part of their overall pricing structure, thus aligning incentives along the chain and increasing the possibility of success as a business (Lundy and Fujisaka, 2008).

Regardless of whether or not the model selected is based on existing actors and skills or is induced, these models can be grouped depending on the focus they accord to diverse actors in the chain. Existing models tend to fall into three general categories: (i) those that focus on developing and supporting producer organizations; (ii) those that focus on specialized intermediaries; and (iii) those that are driven by buyers. Despite the differences in entry points and emphasis, all the models seek to connect actors to facilitate effective market integration.

Producer-driven models

Producer-driven models such as cooperatives and farmer-owned businesses have had a mixed record of providing members with economic benefits in terms of access to dynamic markets. Research in eight countries (Reardon and Huang, 2008) found that membership of producer organizations was correlated with participation in modern markets in only half of the countries; in the rest the correlation was not significant or was negative. This is indicative of the very diverse roles of producer organizations, from political lobbying to providing channels for government subsidies. Marketing cooperatives are rare, and members typically remain oriented to the traditional commodity markets. In cases such as Honduras, where they do exist, agribusiness has been averse to purchasing from cooperatives due to slow decision making and limited entrepreneurial focus (Agropyme, 2006).

But collective action remains an important strategy to increase small-scale producer participation in emerging modern markets and to generate sustained commercial flows of high-quality products. Effective business organization is critical. Economic- and business-focused producer organizations differ from welfare organizations in their entrepreneurial orientation and capacities, and may build on existing informal networks of farmers and traders, as well as support from buyers or other chain actors. Business-oriented cooperatives and employee-ownership models in Europe and North America provide some insights on how this may be achieved but much remains to be learned from existing informal network models common throughout the developing world. An intriguing, if incipient, case is that of Mabeli S.A., a community-owned essential oils corporation in highland Guatemala, where 51% of shares are held by a community development corporation and 49% by producers of the firm's raw materials (Lundy and Fujisaka, 2008).

With regard to organizations that are driven and owned by small-scale producers, such as Cuatro Pinos in Guatemala (Box 4) and NorminVeggies (Concepcion *et al.*, 2006; Box 9) in the Philippines, a rich range of models exists to allow organizations of producers to collectively market despite membership heterogeneity (in terms of land and non-land assets), which can otherwise lead to conflicts of interest within an organization. These management models balance member inclusion and group competitiveness, and involve differentiation of membership to cope with the range of landholdings, wealth, education, etc. These include quasi-membership arrangements and top-up suppliers, or clusters around lead farmers, whereby financially independent growers

---

Box 4. Cuatro Pinos, Guatemala. (From Lundy, 2007.)

Cuatro Pinos is a successful cooperative with nearly 30 years of experience in the vegetable export business. Recently, the cooperative has succeeded in opening large markets for several fresh vegetable products in the USA through an alliance with a specialized wholesaler and several retailers. Existing demand significantly outstrips the capacity of cooperative members, requiring the integration of new producers, organizations and geographies. To achieve this, Cuatro Pinos identifies existing farmer groups, including associations, cooperatives and lead-farmer networks, in favourable environmental niches, works with them to test production schemes and then contracts those that show an ability to meet quantity and quality targets. The cooperative signs a legally binding contract with the producer group, which specifies quantity, quality and a production schedule, as well as providing a fixed annual price for the product. Credit in the form of inputs and technical assistance is provided. This is later discounted from the first few product deliveries. Cuatro Pinos provides business and organizational support to its partner organizations to increase their efficiency and access additional funding from diverse sources for development activities. In 2006, Cuatro Pinos partners successfully raised US$1.7 million for investments in irrigation, packing sheds, education and housing. Through this model Cuatro Pinos has achieved an annual growth rate of 50% in vegetable exports over the past 3 years and expanded from 560 member producers to a network of more than 2000 families. Nearly all the new producers in the network are from regions with higher than national average poverty levels and with limited access to land.

---

create market opportunities for small-scale farmers. Any member differentiation can be a challenge to the cooperative ethos of equality and equity.

Despite the success of Cuatro Pinos and other models, these remain the exception rather than the rule in producer-driven models. Common limitations in farmer organizations include an excessive focus on democratic governance, which, in many cases, leads to effective leaders being replaced every 12–24 months as stipulated by by-laws. This is avoided in the case of Cuatro Pinos by having a professional management team that reports to the elected Cooperative board but is not subject to annual or biannual elections.

Buyer-driven models

Buyer-driven models seek efficiencies in the chain to the benefit of processing and retail. There are some very promising cases where the necessity of organizing supply from a small farm base, often the case with milk procurement, for example, has led to sustained inclusion of small-scale farms.

The classic model is where the buyer integrates backwards and coordinates production (see Boxes 2, 3, 5 and 6). Both the producer and buyer ends of value chains usually want to 'cut out the middleman' and want more competitive buying markets in order to make a shift from a dependency on traditional wholesale markets in pursuit of value, improved quality and product assurance.

Box 5. Dimitar Madzarov in Bulgaria. (From Bachev and Manolov, 2007.)

The private Bulgarian dairy processing firm, Dimitar Madzarov Ltd, has increased by a factor of 20 its daily processing of milk, sourced from over 1000 small farms, half of which have fewer than five cows. The firm has successfully met all the requirements to continue selling its dairy products in a demanding and highly competitive market. Part of the success of Madzarov in building a reliable milk procurement system has to do with the high frequency of payment to its small-scale farmer suppliers. In the case of the smallest farmers, the firm goes as far as advancing payment. Access to this source of timely and reliable financing is considered by the farmers to be of greater importance than the price received for their milk.

Box 6. MA's Tropical Food Processing (Pvt) Ltd, Sri Lanka. (From Samaratunga, 2007.)

MA's Tropical Food Processing (Pvt) Ltd, established in 1987, is a family-owned spice processing enterprise in Sri Lanka, which has shifted its focus to a centralized procurement system. The centralization process has increased the efficiency of procurement through the reduction of the coordination cost.

Procurement is centred on the Regional Agribusiness and Perennial Crop Initiatives and Development (Pvt) Ltd (RAPID), which is responsible for the backward integration of the company's activities in the supply chain and for delivering its social responsibilities to the region. It provides extension services to the farmers on production, record keeping and postharvest practices, organic certification, supply of high-quality planting material and intermediation of commercial credit from banks. It assures continuous supply of raw material at the right time in the right quantity and quality and 'eliminates non-essential intermediaries' from their supply chain. It has resulted in improved information flow among the supply chain segments while reducing the marketing risk faced by both the company and farmers.

The company sets its own private standards, which facilitate the standardization of the products procured from different suppliers and differentiate the company's products from competitors. Further, the company offers farmers a considerable adjustment period to bring the produce up to the standards and pays premium prices to farmers who meet those standards. The company focuses on moving towards logistic improvements in the supply chain by introducing new operations, which have not existed earlier in the areas of grading, processing, packaging, labelling, trademarking, etc. Those practices have made the company more competitive in the local and international markets, enabling its products to satisfy the newly emerging trends in consumer preferences.

Direct procurement is often presented as a win–win–win for customers, business and producers. Improved information flow among the supply chain segments can also help reduce the marketing risk faced by both the company and the farmers. Another reason for businesses to organize their own supply base is the lack of collective action by producers, often due to suspicion of cooperatives or laws that insulate producers from the market by obliging farmers to trade through local government-controlled wholesale markets, such as the APMC Act (law governing the marketing of agricultural products) in many Indian states and the Wholesale Markets Law in Turkey.

Another example of a buyer-driven model from Sri Lanka is the supermarket company Food City. This retailer has a high market penetration for food by South Asian standards (15%), with nearly 120 stores, and a focus on middle- and low-income consumers. Like MA's (Box 6), the management of Food City has a strong commitment to the reduction of rural poverty through its role as purchaser of quality products. The company has made investments in backward linkages (fruit, vegetable, rice and milk) and food processing (meats, ice cream and processed fruits and vegetables). Food City is now looking at regional expansion in Pakistan and Bangladesh.

Given the difficulties faced by producers in Turkey to organize themselves, the few cases of successful direct relations between supermarkets and producer organizations are largely implemented and promoted by supermarkets. For example, Migros Türk achieved direct sourcing with the Narlidere Village Development Cooperative in the Bursa region where others failed, only because of its historical background and its anchoring within the Turkish agrifood chain. Migros invests in capacity building of its supply cooperatives' staff and supports production management, thus going far beyond the incentives comprised of formal contracts (Lemeilleur and Tozanli, 2006).

'Contract farming' can be successfully used by businesses to link small-scale producers to modern markets where capital, technology and market access constitute key limiting factors (Eaton and Shepherd, 2001; FAO, 2008). Contracts provide benefits to traders and processors by removing the risk of periodic shortages and volatile prices, which can be costly if they are servicing large downstream contracts written in advance of a season (Hayami and Otsuka, 1993), or by allowing access to land, which may not be available to expand plantation-scale production. Contract farming can also be an effective mechanism for risk management, because a well-run contract scheme with proven production technology and guaranteed markets can help reduce risks normally faced by unorganized farmers, as seen in the case of Cuatro Pinos in Guatemala in Box 4. For farmers with small landholdings, a contract can also be used as guarantee for loans; there are a growing number of providers of finance, such as Root Capital,[5] who are prepared to provide cash flow credit to smallholders who have secure contracts in place.

Organization of producers is just as important for contract farming as management, and enforcement of contracts with individual smallholders is not viable. Research from the Indian Punjab shows that companies involved in contract farming prefer to work with medium- and large-scale producers to reduce their transaction costs and ensure quality standards (Sharma, 2007). Enforcement of contracts with small-scale producers is a thorny issue, especially when market prices exceed the contracted price. Often, having market 'contacts' with whom

---

[5] Root Capital is active in 29 countries in Latin America, Eastern and Western Africa and Asia (http://www.rootcapital.org/where_we_work.php).

agreements can be brokered with a reasonable expectation of compliance is more important than a legally binding but difficult to enforce formal contract.

Contract farming can be an intermediate step in the commercialization of small-farmer production, as farmers innovate and reconnect with more traditional system of brokers, but on their own terms. This, for example, is the case with potato production in northern Thailand (Wiboonpongse *et al.*, 2007).

Models of intermediation

Integrating forward (for producers) or backward (for retailers or processors) is time-demanding and expensive. Business models transferred from the elite retail-driven chains may be as inappropriate for agribusiness as they are for small-scale producers. Despite the attractions of 'cutting out the middleman', organizing direct procurement can have high transaction costs for private players, and have mixed outcomes. In Mexico, Wal-Mart recently tried to buy strawberries direct from the farmers, but withdrew due to high costs (Berdegué *et al.*, 2008b). Given these costs, a business model that works with chain intermediaries, either traditional or new, can offer the opportunity to be profitable in highly competitive, price-sensitive markets.

It is much easier for retailers setting up in emerging economies to procure from traditional wholesalers, and leave the wholesaler to grade for physical quality, unless there are strong market incentives for retailers to guarantee product quality, consistency, safety and traceability. This explains the relative scarcity of evidence of farm-level restructuring and the type of model described in Box 1 in developing and emerging economies (Reardon and Huang, 2008). In Chinese horticulture, where the market is characterized by 50 million autonomous producers, selling on spot terms through five million small traders, where the retail market is very competitive and few companies are making money, and where the majority of customers are not willing to pay for top-class produce, the economics of backward integration are particularly daunting. Although many supermarkets profess to be putting vertical coordination in place, the majority of trade is via traditional traders.

There are, however, some very promising models of upgraded or new intermediaries that are introducing food safety, consistent quality, year-round supply and innovation, at a competitive price. Private companies are emerging as important intermediaries that enable small-scale farmers to supply to supermarkets, as exemplified by Bimandiri in Indonesia (Box 7) and Hortifruti in Honduras (Box 8). Another example is the production network for hot peppers managed by the export firm, Hugo Restrepo and Company, in Colombia and Peru. Under this model, the firm provides services to farmers, such as access to seeds, drip irrigation technology and technical assistance, as well as a guaranteed market via contracts to participating producers, producer organizations and clients involving quality control and guaranteed volumes (Ochoa and Lundy, 2001). While this is not an exclusively smallholder model, it shows the range of services that a specialized intermediary organization can provide.

Box 7. Specialized wholesaler: Bimandiri in Indonesia. (From Sandredo, 2006; World Bank, 2007a.)

The Bimandiri company in Indonesia, which has changed from a traditional wholesaler to a supplier of vegetables and fruits mainly to Carrefour, is an example of a specialized intermediary. Bimandiri encourages farmers to cooperate in producer organizations and works with those groups on the basis of agreed quantities. The company has worked closely with its producer organizations, supplying technical assistance and credit, in order to assure quality standards and consistent volumes for its retailer client. Bimandiri has maintained preferred suppliers lists but moved away from a close extension role. It continues to implement transparent negotiated producer prices.

Box 8. Lead-farmer networks with Hortifruti Honduras. (From Agropyme, 2006; Lundy, 2007.)

Hortifruti is the specialized wholesaler for fresh fruits and vegetables for Wal-Mart in Central America. The company works with a variety of suppliers for vegetables in Honduras and Nicaragua, often purchasing product from existing farmer cooperatives. However, it has experienced significant difficulties with these farmer organizations in terms of lengthy decision-making processes. As a result, Hortifruti Honduras has developed and promoted a 'lead-farmer' model of organization through which it identifies and builds the capacity of farmers who can meet its quality needs in a consistent fashion. After demonstrating such capacity, lead farmers receive larger and larger orders for product or new products and are invited to work with neighbouring farmers to meet this demand. Lead farmers provide access to technology, technical assistance and market access to their network of neighbours as part of a bundle of production and marketing services. The cost of these services is recouped via the sales margin to Hortifruti. The expansion of this model depends on the identification of new lead farmers. Early results indicate that it is low-cost, scaleable and sustainable.

As is clear from the examples of Bimandiri, Hortifruti and Hugo Restrepo, models of intermediation include a strong dose of service provision, including finance – usually by the intermediary organization or specialized providers – to balance both the needs of small-scale farmers *and* the realities of emerging modern markets in terms of quality and volume. For example, the Los Angeles Salad Company, which works as a wholesaler between Cuatro Pinos in Guatemala and the retailer Costco in the USA, not only helps to market products and provides logistics support in the USA, but also provides technical assistance in product quality, access to innovations in packaging and packaging technology and assistance in new product development. LA Salad also helps facilitate production planning and manages over- or underproduction in coordination with other producer regions. Without the provision of these services, the ability of Cuatro Pinos to sell consistently to Costco would be much lower. These new intermediaries are characterized by increased knowledge management (to improve chain coordination and quality), closer links to buyers and incentives for product and process upgrading. This can be an important new role for NGOs, though there is a growing appreciation of the efficiency benefits of upgrading existing intermediaries.

A new generation of commercial intermediary in India is demonstrating that service provision can itself be a profitable part of the business model, which can trigger inclusive growth. The rural retailer, Hariyali Kisaan Bazaar, which is part of the DSCL conglomerate, sells agri-inputs and consumer goods through its chain of centres, which also serve as a common platform for providers of financial services, health services, etc. The Haryali centres are procurement hubs for farm outputs, providing buyback and warehousing (Bell *et al.*, 2007b; Gupta, 2008), and thus creating multiple revenue streams based on transparent and effective participation in input as well as output value chains. Each Hariyali store has a catchment radius of 20–25 km, comprising about 15–20,000 farming families. They aim to provide producers with 'urban amenities in rural areas', easy availability of quality products at 'city-like' fair prices and, through IT, provide commodity prices and commodity futures, as well as ATM access and weather forecasts. On the procurement side, they create linkages between producers and processors, exporters and retailers.

There are examples of producer organizations adding their own commercial intermediary, in the form of consolidation and marketing units (Box 9).

Box 9. Normincorp in Mindanao, Philippines. (From Concepcion *et al.*, 2006.)

Farmers of the Northern Mindanao Vegetable Producers' Association, NorminVeggies, are able to successfully participate in dynamic vegetable chains primarily because of the organizational structure they chose in order to respond to the market challenges. This involves a corporation, Normincorp, which gives them the agility needed for each development in the supply chain. Normincorp's formation signified a new development in marketing for small farmers. While established as a stock corporation, Normincorp functions more like a cooperative and has a social enterprise character. It was set up and operated with a keen business sense, and also with full empathy for the small farmers. As market facilitator, Normincorp saw to it that production was programmed by farmer clusters with their respective cluster leaders, according to marketing plans, that quality farm and postharvest management could be done by each farmer in the cluster, and that coordination could be provided for the sequence of activities that include order taking, outshipment logistics, billing/charging, collection and remittance to the farmers. For these services, Normincorp earns a market facilitation fee based on the value of the sale and uses the income to cover the marketing management overhead.

Normincorp is not a trading company. Rather, it is a market facilitator linking the farmer through his or her cluster directly to the buyer. The farmer is given the buyer's price, and he/she is therefore accountable for the product and retains ownership of the product up to the point of sale. This encourages the farmer to supply the best quality since the price is given to him/her and all sales are remitted directly after deducting the market facilitation fee, which is based on the quantity of accepted vegetables. Conversely, all rejects are individually charged to the concerned farmer. Labelling of products per farm or farmer provides this traceability.

Working with this new generation of 'doubly specialized intermediaries' (which are both business-oriented and development-motivated) such as Normincorp is an area that appears to offer the greatest potential for linking large business with small-scale producers.

Much more common at present are market-oriented but traditional traders taking steps to improve quality in their supply chains, where suppliers produce to the traders' specifications (crop management, harvesting, packaging, etc.), and where the traders invest in supplier training and other investments. A very interesting example of a butterhead lettuce supplier to Ho Chi Minh City in Vietnam has been identified by Cadilhon (2006). The farmer collectors who supply the intermediary train farmers to grow and harvest high-quality lettuce. Through this collaboration, and through investments and forward planning with regular suppliers, the intermediary only gets high-quality product. In China, agricultural brokers and traders were denigrated for several decades and the government tried to ban them, but without success. The government realized the vital role that an agricultural brokers' association can play as a bridge between small farmers and outside markets, and in contributing to farmers' incomes and rural development. It therefore adopted a new strategy designed to organize them and regulate their activities after the economic reform (Shudon, 2008).

Export-oriented companies setting up in new supplier countries almost always rely on intermediation to simplify decision making, reduce risk and lower transaction costs.

Alternative trade models

Alternative trade models cover a range of initiatives that make use of third-party certifications to monitor compliance with selected indicators valued by diverse members of the supply chain. Chapter 7 (this volume) looks at this issue in depth, particularly at how standards line up with issues related to CSR. Alternative trade models can be divided by their principal focus. In the chapter on CSR, the authors identify four categories based on the principal goals of the standard: (i) environment; (ii) social; (iii) benefits to the local economy and community; and (iv) food safety and quality. For the purposes of this chapter, we will briefly discuss standards that seek to promote benefits to the local economy and community and how they seek to resolve the issue of business models. We use Fairtrade as a model of what is, admittedly, a much wider pool of alternative trade standards.[6]

Of the existing alternative trade models, perhaps the best known is the Fairtrade movement, which has the objective of creating opportunities for

[6] Other relevant standards that speak to small-farmer viability include SCS-001, Basel Criteria for Responsible Soy, Rainforest Alliance, the Roundtable on Sustainable Palm Oil and the SAI Principals and Practices for Sustainable Production. However, many certification schemes do not contain elements of business models which may be considered as central to inclusion of small-scale producers, such as transparency (including transparency of how the certification premium is allocated), collective action, durability of trading relationships, etc.

economically disadvantaged producers as its strategic intent. Is there then, in Fairtrade, a shortcut to inclusive markets? Is Fairtrade a valid business model for large companies to translate into their mainstream trading, or at least a source of elements for new business models?

Fairtrade has at its core the concept of 'fairer' pricing that gives growers in developing countries a better price for their work and gives longer term stability to producer–buyer trading relationships. The umbrella organization Fairtrade Labelling Organizations International (FLO) stipulates two sets of generic producer standards, one for small farmers and one for workers on plantations and in factories. The first set applies to smallholders organized in cooperatives or other organizations with a democratic, participative structure. The second set applies to organized workers whose employers pay decent wages, guarantee the right to join trade unions and provide good housing where relevant. On plantations and in factories, minimum health and safety as well as environmental standards must be complied with, and no child or forced labour may occur.

As Fairtrade is also about development, the generic standards distinguish between minimum requirements, which producers must meet to be certified as Fairtrade, and progress requirements that encourage producer organizations to continuously improve working conditions and product quality, to increase the environmental sustainability of their activities and to invest in the development of the organizations and their producers/workers.

The standards stipulate that traders have to:

- pay a price to producers that covers the costs of sustainable production and living;
- pay a premium that producers can invest in development;
- partially pay in advance, when producers ask for it;
- sign contracts that allow for long-term planning and sustainable production practices.

Finally, there are a few product-specific Fairtrade standards for each product that determine such things as minimum quality, price and processing requirements. These have to be complied with.

Some elements of the Fairtrade model – such as commitment to long-term trading relationships – are cornerstones of inclusive business. But, as a business model for wider application, there are a number of limitations, some of which are easier to resolve than others. There are deep tensions between the smallholder and plantation standards, and the lack of a clear definition of 'disadvantaged producers and workers' based on access to markets as well as income. In contrast to pricing models which are based on a combination of market conditions and product quality, Fairtrade floor prices are fixed by FLO, and have proven slow to change even in sectors where significant market price increases have occurred, such as coffee.

For many large companies, Fairtrade accounts for a small proportion of their overall purchases. Many food retailers have positioned Fairtrade as an up-market niche, as a test of their customers' willingness to pay for 'non-exploitative' trading with primary producers, rather than as a *corporate standard* and a means to transform their mainstream businesses (Tallontire and

Vorley, 2005). Even within the Fairtrade movement, questions are being asked about whether the purchase of certified Fairtrade goods is an effective way of achieving systemic fairness in trade. These groups are investing in other approaches such as in schemes to facilitate improved access to conventional markets for marginalized producers, and lobbying on codes of practice on retailer–supplier trade relations.

How Are These Models Impacting on Smallholders?

Impacts of different business models on smallholder farmers vary depending on the model employed and the way this is implemented by the chain actors. As the business axiom states, 'you can't manage what you don't measure' and increasing attention is being paid to the development of tools to assess the impacts and sustainability of diverse business models. The paper on *Inclusive Business* by the World Business Council for Sustainable Development and SNV Netherlands Development Organization (WBCSD and SNV, 2008) distinguishes between direct economic benefits, indirect economic benefits and broader social benefits. Recent academic work highlights considerable attempts at developing metrics for supply chains but none has been fully implemented (Aramyan *et al.*, 2006). The area of key performance indicators for supply chains remains under development and is limited by a lack of transparency and cooperation among supply chain actors (Van der Vorst, 2006). Despite these limitations, it is clear that business models have both quantifiable and qualitative impacts on smallholder farmers. Both categories need to be measured and assessed to understand the effects of the business model on rural populations.

The first category comprises key *quantifiable indicators*, which are focused on measuring chain-wide evolution. Those indicators relevant to smallholders include production volumes, product quality, net income, distribution of income among smallholders, within households and along the supply chain, as well as the distribution of costs associated with risk mitigation and management. These indicators can be complemented with additional quantitative measures that assess the overall 'health' of the supply chain, such as market position and penetration, profitability as compared to similar chains and trends in volume and prices.

A second critical area of impact assessment focuses on *qualitative or skills-based indicators*. While difficult to quantify, advances in skills and relationships underpin and sustain gains shown in quantifiable indicators such as income and profit. Key skills related to the *quality of the trading relationship* focus on negotiation, the construction of sustainable commercial relationships and the governance functions of the chain itself. For chains linked to dynamic market segments, additional attention should be paid to issues related to product and process upgrading and collective innovation as the chain adapts to increasingly demanding market conditions. While this process does not occur fully at the farmer level, the existence of this skill set is critical for continuing competitiveness of the overall system. Unlocking innovation and opportunities for smallholders is a critical element of impact since this leads to benefits that help drive farmer incentives for inclusion.

Changes achieved through *producer organization* models can improve negotiating skills and enhance access to service provision. Models of producer-driven vertical integration – either becoming co-owner of a supply chain or one of its segments in pursuit of value-added – can make sense when built on a business mentality. However, this downstream ownership route may not always compare well to investments in building a network of specialized actors to achieve similar goals. For instance, research in Africa provides interesting evidence in this regard, showing that many of the benefits achieved by relatively autonomous smallholder-owned and managed cooperatives can be captured by more dependent, i.e. less highly trained and skilled, groups if appropriate links are developed with other market actors (Stringfellow *et al.*, 1997). The difference between choosing a strategy of vertical integration versus horizontal cooperation may, therefore, boil down more to costs (money and time) than notably different outcomes.

Some arrangements to sustain the inclusion of all members while maintaining the competitiveness of the organization, such as lead-farmer clusters, which accept differentiation within an organization based on assets, are a significant and challenging departure from the original cooperative ethos of equal treatment for all members. Cuatro Pinos (Box 4) exemplifies this tension. Despite the success achieved in expanding a supply network by nearly 400% while maintaining high product quality, the final distribution of benefits is still skewed towards cooperative members as opposed to non-members, even though non-members provide up to 80% of product volume in some categories. Farmers who do not happen to live in the seven communities where the cooperative was founded 30 years ago are not allowed to become members. This means that they cannot access profit-sharing mechanisms developed by the cooperative, which provide significant additional income to producers at the end of the year. Cuatro Pinos' management is aware of this issue and is seeking to resolve it through service provision and increased prices and/or volumes from non-cooperative members. However, it is members themselves who would need to reform their organizational rules to allow the inclusion of new members (Lundy *et al.*, 2006). To date, this has not happened.

Models focused on *intermediation* drive change through processes of negotiation among actors. They achieve efficiency gains through greater organization along the whole chain through improved information flow and shared standards. The distribution of additional benefits along the chain needs to be negotiated and care exercised in not allowing the intermediary actor to extract additional benefits based on information asymmetries. The development of transparent pricing mechanisms is an important tool to reduce this risk. For example, in Indonesia, the specialized wholesaler Bimandiri (Box 7), which supplies fruits and vegetables to Carrefour, operates a transparent margin system to cover its participation in the system. All actors know the final prices and the intermediary margin, thus avoiding windfall profits for the intermediary organization when market conditions improve and providing an incentive to increase volumes. In other cases, prices are set based on crop models on a yearly basis. This can be done with producer participation, such as in Cuatro Pinos in Guatemala, or a non-participatory fashion, such as in the case of Hugo Restrepo and Company in Colombia. Regardless of how prices are set, clarity on how prices reflect production costs, relative risks and returns is critical to assure greater equity along the chain.

Buyer-driven models affect smallholders through the application of (often strict) norms and standards relating to quality and volume. They tend to push processes of functional improvement up the supply chain, often with limited incentives to compliance beyond continued participation in the market. Because of their proximity to the end buyers, these models can identify consumer trends and can provide clear incentives for market-driven product and process upgrading. Additional benefits tend to accrue to buyers and care should be taken to achieve transparent assessments of gains and meet equity concerns.

Cases of inclusion driven by private businesses are characterized by small farmers having less say in the governance of the chain and by less capacity building of small-scale suppliers beyond production and postharvest management. Where a buyer organizes a network of producers from a corporate responsibility ethic, the risk is more one of paternalism and dependence. On the other hand, case study analysis (Berdegué *et al.*, 2008a) found no evidence that in such situations small farmers will have lower direct economic benefits, at least in the short run. Also, under these conditions small farmers do not need to incur the costs of coordination or of collective action. In the case of MA's in Sri Lanka, there were clear income benefits for smallholder suppliers (Box 10).

A buyer-driven network can be managed through using the transparent pricing strategies highlighted above as well as the incorporation, where possible, of incentives based on quality. For example, the US speciality coffee company Intelligentsia Coffee and Tea manages a 'direct relationship' quality-based model with producers. This model prices coffee based on its cup quality, with payment going directly to the producer. Additional services needed to move coffee from the farm to the US market are contracted by Intelligentsia directly and not discounted from the farmer price (*New York Times*, 22 June 2006).

A drawback of buyer-driven models for producers is the frequent demand for exclusivity. From a processor or retailer perspective, a supply chain is a source of competitive advantage, and they will seek to exclude competitors and prevent suppliers from 'side-selling'. Because a buyer has invested in the supply network, and because the buyer needs to be able to fulfil contractual obligations for specific volumes to *its* customers, it will demand exclusivity from its smallholder suppliers. This can be frustrating for producers, who do not see transparency in how prices are set, or in how quality discounts are often determined. The Kenyan supermarket Uchumi makes a point of not demanding exclusivity from its smallholder suppliers. Another way around this issue is to have prices set weekly rather than fixed at the start of the season, to reduce discrepancies between contract and market prices (Box 11).

Alternative trade models, especially Fairtrade, have demonstrated success in benefits transfer and consumer acceptance. None the less, an important percentage of the Fairtrade premium resides with certification and coordinating agencies. Gross margin at retail level is much higher than at other levels of the

---

Box 10. Impact assessment of MA's procurement system, Sri Lanka. (From Samaratunga, 2007.)

An impact assessment of the smallholder procurement system established by MA's Tropical (Box 6) showed clear improvements in corporate income, volume of trade, assets, farm income, employment creation and non-monetary benefits while ensuring a greater degree of inclusion of small farmers in the new supply chain. With inclusion in the company's supply chain, the farmers achieved a premium price for better quality products, price stability, a spread of income throughout the year, and services such as extension, credit facilities and marketing risk minimization.

Average yearly per acre income comparison (Rs/acre/year)

Farmers supplying to MA's	98,000
Non-supplying farmers	48,000

Price comparison between MA's and village trader (Rs/kg)

	Cinnamon	Nutmeg	Cloves	Pepper	Citronella
MA's	675	300	580	180	22
Village trader	550	150	360	125	12

This model has been in existence as a sustainable system for about a decade while increasing its capacity for greater inclusion of farmers. The company has not yet reached its potential capacity.

However, there is some evidence of exclusion through the higher transport charges, delayed payments, use of cheques as the mode of payment and the low production capacity of the company. Even though the company has initiated an informal farmer organization, lack of coordination and poor structure have excluded a certain stratum of farmers from the system. Lack of quality consciousness and credit-bound relationships with the village traders also have some form of correlation with the exclusion of farmers from the chain.

Greater inclusion can be attained by creating an arrangement for the transport cost to be borne by the company or farmer organization and by introducing a liquid mode of payments.

Box 11. Adjusting payment terms: Vegpro in Kenya. (From Bell *et al.*, 2007a.)

In common with other leading exporters of fresh vegetables from Kenya, Vegpro divides production between its own farms and smallholder outgrowers, which it relies on for crops that are not well suited to plantation production, such as peas. In 2007, Vegpro was purchasing most of its snow peas from 3500 smallholder farmers organized into 50 self-help groups. Despite the coordination offered by the self-help group structure, it is no small task to ensure consistent volumes, quality and standards across 3500 farmers. Vegpro had previously been paying farmers a fixed year-round price that exceeded the average market price over the course of the year. When the market price was below the fixed price, farmers had been content to sell to Vegpro, submitting volumes that apparently included uncertified produce from their neighbours. But, when the market price rose, farmers in need of cash would side-sell to local traders. Vegpro reduced side-selling by employing field supervisors and switching from annual fixed prices to weekly prices set in relation to the market price.

value chain. Consequently, it has been argued that consumers of Fairtrade products are supporting the shareholders of the international retailers more than the actual smallholder target groups. Incentives for product improvement and innovation have been traditionally weak, with limited feedback regarding consumer trends and demands beyond that covered by certification.

None of the above business models is inherently superior for smallholders, with the possible exception of classical Fairtrade. To initiate processes quickly, intermediary or buyer-driven models are useful as they provide turnkey solutions and somewhat lower risk. In the medium term, however, the promotion of stronger producer organizations may build greater resilience and increased participation in chain governance, especially if combined with specialized intermediaries.

The selection of a specific model or elements from various models is highly dependent on market conditions, participating actors and their knowledge and skills and the existence (or not) of support agencies and policies. As market linkages evolve, models need to adapt to respond to changing market conditions as well as in the relationships between the participating actors. Approaches need to be piloted for specific locations, products, conditions and markets in order to better understand how to update current models, and which forms of best practice need to be adapted to help support sustainable impact for smallholders.

What Can Be Done to Help Prepare Smallholders to Participate?

What needs to happen at the farm level and in supply chains to support the participation of small-scale farmers in dynamic and more profitable market segments, from a business model perspective?

According to a study of 35 successful farmer-owned rural businesses in Latin America (Camacho *et al.*, 2007), producer organizations follow a similar trajectory of skill development that includes capacities focused on: (i) market linkages for goods and services; (ii) increased internal and bridging social capital; and (iii) the development of professional management capacities. The development of these skills requires access to effective business support services, effective alliances with other chain actors and an effective enabling environment.

Support services may be technical, managerial or financial in nature, provided by diverse types of formal and informal service providers. But they share several common factors: (i) a focus on effective solutions to bottlenecks that cause exclusion; (ii) a business orientation to guarantee sustainability over time; (iii) flexibility linked to client needs; and (iv) provision by operators close to the clients. The topic has been covered in depth in the Business Development Services (BDS) literature.[7] For many smallholders, service provision between commercial actors, known as embedded services, holds promise in that these services depend on commercial incentives rather than public subsidies.

---

[7] www.bdsknowledge.org

Financial services are crucial for farmers to access dynamic markets and sustain their participation in them. As supermarkets and processors tend to pay only after a certain period (often 45 days or more), there needs to be a mechanism to bring liquidity into the supply chain. In addition to working capital provision, other financial services such as cash flow finance, in which the commercial relationship rather than collateral assets guarantee the loan, can be arranged as three-way agreements among buyer, producer and finance institution. The informal moneylender is likely to guard customer information on risk and viability. One option is to transform the moneylender into a bank worker. This has been proposed by some NGOs and external funders, but there has been patchy success with this. A version of this transformation rewards knowledge and innovation by the moneylender through private sector arrangements.

There are often good reasons for a lack of embedded services in rural areas. Successful services rely on knowing the customer well and research shows that this knowledge is difficult for non-residents to obtain or interpret. The use of tools that strengthen the capacities of informal service providers to identify, provide and improve their embedded services in rural areas is of critical importance.

A crucial point in times of high and volatile market prices is the development of models of *reciprocal responsibility* between buyers and producers. As prices and demand rise, producer organizations will be tempted to break contracts and side-sell committed volumes to other buyers offering higher prices. This might generate additional income in the short term, but it is critical to recognize that sustainable relationships can generate additional negotiation power in the long run. In addition, sustainable relationships lay the groundwork for the development of joint ventures for new product development and co-investment. Opportunistic behaviour works directly against this possibility.

The above interventions must link to a suitable enabling environment, the components of which are described in more detail in the section 'What are the Priorities for the Public Sector?' and in Chapter 5 (this volume) as well.

What Do Business Partners Have to Consider and Do in Order to Work Successfully with Smallholders?

We have seen in the first section that the biggest challenge for large businesses to work with small-scale farmers is that of organizing supply. Without a means to reduce transaction costs, ensure due diligence and ensure that trading agreements are honoured, they will see smallholder suppliers as a threat to their 'value proposition'. In the section 'What Is the Business Case for Adjusting Business Models in Favour of Smallholders?' we saw that there are many examples of companies organizing their own supply base and setting up producer groups, especially where there is a lack of collective producer action. But organizing direct procurement is costly for private players, and such efforts are likely to remain as small CSR pilot projects. Where there has been positive business action, it has largely focused on niche export markets to the north rather than the much more pressing challenge of inclusive development *within* 'transforming economies' where 80% of the world's rural poor live.

Opportunity lies in the 'Partner network' part of the business model template in Figure 1. Much private sector policy is rooted in the procurement-profit philosophy, without extending this approach to co-investment or partnership win–wins. A suggested refocusing of private sector innovation and incentives on sustainable supply from small-farmer networks has the potential to unleash the best of all worlds.

Upgrading mainstream procurement

Much can be done by businesses to upgrade mainstream procurement within their existing model to ensure that their procurement practices work to the benefit, rather than detriment, of small-scale producers and suppliers. The clear business incentives are continued access to supply, option to be the 'buyer of choice', access to better quality supply and a social licence to operate. Points of focus here are coherence between corporate policies and actual procurement practices, through adjustment of reward systems for buyers, and through senior management buy-in. A reorientation of training and development awareness of buyers is a first priority. The asymmetries of market power between sellers and buyers have, until recent food price rises, allowed retailers to simultaneously extract both lower prices and higher standards from suppliers. Traidcraft's reports on purchasing practices are an excellent source of further information (e.g. CIPS and Traidcraft, 2008). Payment terms and contracts can be adjusted to the realities of smallholder production without compromising commercial imperatives (Boxes 11 and 12). Buyers are quick to criticize 'side-selling', but may readily engage in 'side-buying', procuring opportunistically outside of established supplier networks for short-term profit.

Frequent and consistent access to information on market trends, projected volumes and production technology, in addition to shared decision making in regard to chain rules and price structures, is also critical. Better forecasting and planning can reduce some of the pressures on suppliers that drive poor working practices and casualization of labour.

Box 12. Adjusting contracts: Postobon in Colombia. (From Espinal *et al.*, 2005.)

One way to handle the problem of side-selling is through specific agreements that recognize opportunities and include them openly in negotiations. In Colombia, demand for tropical fruit pulp exceeds supply. As a result, the private sector firm Postobon began offering annual contracts to smallholder blackberry farmers that contained two market condition-related clauses. In times of high market prices (a seller's market), producers were allowed to sell up to 20% of their total volume to other buyers principally for the fresh market. In times of low market prices (a buyer's market), Postobon was allowed to purchase up to 20% of its total volume from non-contracted suppliers. These agreements explicitly recognized the pressure for opportunistic behaviour and identified mechanisms to manage them.

There are also models of *inclusive procurement*, built on preferential sourcing from small-scale producers and family farmers and their organizations. For example, Carrefour Indonesia has established a dialogue with SME suppliers of fresh food (vegetables, fish), household equipment and textiles, to improve product quality and packaging and improve their shelf access, in part by waiving the listing fee normally charged to companies waiting to sell to the chain. Similarly, Wal-Mart, Honduras, has established the 'Una Mano para Crecer' ('Help to Grow') programme for SMEs.

Part of a commitment to inclusive procurement should look at alternatives to paternalistic supply systems and demands of supplier exclusivity. While it is tempting to want to 'cut out the middleman', chain intermediaries often are vital in linking smallholders to dynamic markets, and are of particular importance to the poorest farmers and to those located further away from the markets and the main roads. There is much for food processors and retailers to do to cultivate efficient intermediaries, including those set up through producers' own initiatives, rather than seeking to eliminate them from the chain.

Better standards

The issue of private sector standards is also central to pro-smallholder business models. GLOBALGAP is now a passport to the most demanding export markets, but many compliance costs are not a function of the volume of production, but are per-farm costs, thus pushing up the per unit cost of compliance for small-scale producers. This applies as much to standards for 'sustainability' as to those for food safety and traceability. A lack of coordination between schemes means farmers certified to multiple standards must pay for separate audits. There are pro-smallholder approaches to standards, including group certification and combined audits, as well as the use of local certification agencies, but also more fundamentally the participatory development of standards, involving the farmers who will have to implement them – the 'standards takers' – from the outset.

Pan-industry initiatives

Not all aspects of business models are competitive. Much can be achieved through industry collaboration to create an environment for more inclusive markets. Cross-industry codes of conduct established by the business sector and regulated by them, for example, in Argentina (Box 13) can provide much needed oversight of trading relationships at the domestic level.

What Are the Priorities for the Public Sector?

Innovative business models can make a positive difference in terms of inclusion. The role of public policy is not a primary focus of this chapter. But it is important to note the potential of proactive policies – including infrastructure,

---

Box 13. Best commercial practices code in Argentina. (From Brom, 2007.)

Rapid investment by global and regional retail players in Argentina in the late 1990s created fierce competition with local retail investors creating a trading environment unsatisfactory to small companies, a poor bargaining position for many and complaints at all levels. The choices faced by the sector were either to develop a private code or to submit to government legislation. The Food and Beverages Manufacturing Association (COPAL) and the Argentine Supermarkets Chamber (CAS) worked together with reference to evidence and experience from across the globe to develop a private code of practice, which was signed in June 2000. Since then supplementary rules have been added and the approach shared with many countries in the region and indeed worldwide. Similar private sector codes have, for example, been developed and adopted in Colombia and Mexico. Seven years on, there has been significant improvement in both free and fair practice and thus competitiveness. The culture and way of doing business have changed with a dramatic decline in cases submitted for mediation or arbitration.

---

finance and support services – to stimulate and support those types of business models which are more inclusive and that are also good business. There is a vitally important role for the public sector to facilitate successful alliances between smallholders and larger business, especially if successful small initiatives are to be scaled up.

The enabling environment[8]

A priority area of intervention is that of the enabling environment. Recent work by the World Bank on agricultural innovation systems identifies a range of options to support an enabling environment that promotes innovation. Key findings from this work include the importance of using targeted public and private research investments to resolve technological bottlenecks in the supply chain, the inclusion of social and environmental sustainability criteria, a focus on outcomes in terms of poverty reduction and a focus on collaboration among actors as a driver of competitiveness (World Bank, 2006). The consistent provision of key infrastructure services (roads, water, electricity and communications) is a central element of an enabling environment, as are relevant public policies to maintain a competitive market, oversee the working of contract laws and contractual enforcement, and oversee FDI and taxation.

Investment in traditional and wholesale markets is clearly an important priority for public policy. Where wholesale markets fail to keep up with changes in retail – especially the supermarket revolution – they can fall into decay. Traditional markets can be a bridge for small-scale farmers to increase their capacity and to eventually link to modern markets. Successful upgrading and modernization of wholesale markets and their procurement networks also

---

[8] See also Chapter 5, this volume.

require upgrading and modernizing of their primary clients – the traditional retail sector – if they are to remain crucial players favouring inclusion of small-scale producers.

Where land is unequally distributed, as in South Africa, this becomes a significant determinant of market inclusion as the modern market will always seek to source from the large farm sector. Under these conditions of dualistic farm structure, inclusion attempts will be working against gravity, and public policy has a vital role.

Donors are increasingly interested in facilitating the bridge between the majority of small-scale producers and modern markets. Businesses can develop effective initiatives in partnership with governments, donors and NGOs, and can learn as much from the successes and failures of development agencies and NGOs as the latter can learn from business. For example, the self-service wholesale operator METRO Cash & Carry is working with the Vietnam Ministry of Trade and the German development agency, GTZ, to support development of Vietnam's distribution network. However, until these donor-supported initiatives are scaled up and become self-supporting, the question of tokenism and long-term sustainability remains. As an alternative approach, a number of donors have in place business challenge funds. The Africa Enterprise Challenge Fund, the Financial Deepening Challenge Fund and the USAID Global Development Alliance offer the opportunity for innovative business models within inclusive agrifood markets to be both explored and developed.

At some point governments must balance equity and efficiency, despite the compelling case to support the huge numbers of small- and micro-scale farms. The costs of inclusion and exclusion must be evaluated in considering policy options. Evaluation of future scenarios should attempt to include estimates of these costs in order to provide additional insights to the real costs and benefits of the policy options. Case study evidence suggests that inclusion into restructured markets may be unsustainable for the 'poorest of the poor'. There is a lack of data to inform resource allocation and thus, for example, the threshold for support. Such thresholds can be a minimum size of farms, but may also include non-land triggers such as completion of training or membership of a producer organization.

Partnership facilitation and chain-wide learning

A key pattern in successful linkages between small-scale farmers and dynamic markets is the collaborative arrangement between: (i) trained and organized farmers; (ii) a receptive business sector; and (iii) conducive public policies and programmes (Figure 2). Such arrangements may benefit from specialized partnership facilitation. Innovation in building inclusive markets is greatly enhanced when business actors within the market chain engage along the whole chain, together with indirect businesses (input suppliers, etc.), and with relevant public institutions. If interventions are made without coordination, they can lead to market distortions instead of market development, potentially flooding markets and supporting inefficient production systems.

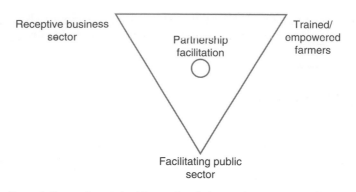

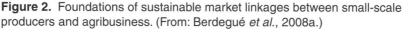

Figure 2. Foundations of sustainable market linkages between small-scale producers and agribusiness. (From: Berdegué *et al.*, 2008a.)

   The development of collaboration among actors requires linking actors in ways that facilitate discussions and information exchange among them. Examples of how this can work include chain-wide committees facilitated by Ministries of Agriculture, 'inter-professional' or commodity associations formed by the chain partners (Shepherd and Cadilhon, 2009), and the use of public–private partnerships. Difficult issues such as power and knowledge asymmetries need to be carefully managed to avoid excluding weaker members of the chain. Examples include work in Colombia[9] and Honduras, as well as a link between the Centre for Agricultural Policy and Agribusiness Studies (CAPAS) at Padjadjaran University in Indonesia and Carrefour Indonesia. A memorandum of strategic cooperation between the two involves developing a supply chain model involving small-scale suppliers, development of new agricultural products, transfer of know-how and channelling of products to the Carrefour quality line programme.

Encouraging procurement from small and family-scale farmers

   Apart from providing an enabling environment and appropriate services, there are examples of specific policy innovations to encourage procurement from small and family-scale farmers:

- The biofuels Social Seal in Brazil is a promising example of a tool to improve the equity of the 'biofuels revolution' by providing the downstream biodiesel industry with incentives to source their feedstock from smallholders and family farmers (Abramovay and Magalhães, 2007).
- Private–public partnerships in Michoacán, Mexico, have been organized to coordinate the production and marketing of avocados built around phytosanitary standards (Medina and Aguirre, 2007).
- The use of policy pressure or incentives to agribusiness and retail for pro-poor procurement, such as requiring supermarkets to provide adequate space in their shelves for small-scale farmers' products.

[9] http://www.agrocadenas.gov.co.

- A supplier ombudsman with an independent regulatory role to oversee the way in which powerful buyers such as supermarkets engage with their suppliers has been established in Australia.

Closing Comments

There is a sound business case for securing and enhancing small-scale producers' inclusion, which can bring both economic and wider development gains. This requires that appropriate business models are applied and, where applicable, that this is done in partnership with producers, the public sector, intermediaries and development agencies. Two big challenges are evident when seeking to apply inclusive models to developing country economies dominated by small-scale producers, either for domestic retailing and processing, or for exporting. The first is organizing and upgrading supply from a dispersed producer base. The second is traceability and quality assurance. Through attention to the 'partner network' in the business model framework (Figure 1), the value proposition of modern agrifood business and cost structure can be maintained or even strengthened by building in inclusion of small-scale producers and suppliers.

Successful models tend to evolve towards a common set of principles (Hobbs *et al.*, 2000). These include: (i) greater information and knowledge flows; (ii) a focus on differentiated products; (iii) an orientation towards market demands; and (iv) chain-wide organizational structures that recognize the interdependence of actors and facilitate collaborative problem solving. The sum of these principles is *systemic competitiveness*, which is based not only on the efficiencies of individual actors but also on collective efficiencies. The classic business model schematic in Figure 1, which describes the individual firm, therefore needs to be revisited to reflect how chain actors can collaborate to build a chain-wide model that balances risk, responsibilities and benefits along the chain while not undermining competitiveness.

The business models concept is especially useful in helping business to understand the reach of downstream decisions on how value is created or lost by supply chain actors, including smallholders. But business models that work for more inclusive market development are not exclusively about procurement. The approach compels us to look to effective alliances and linkages by all chain participants. This rarely occurs spontaneously, given the often adversarial relationships that characterize commercial links in the agrifood sector. As a result, specific actions to clarify and develop plans for collective action at the chain scale are needed. Some good tools can be found in participatory chain analysis and upgrading manuals (Lundy *et al.*, 2006; Vermeulen *et al.*, 2008).

Another benefit of the business models concept is that it forces us to rethink CSR. The contemporary approach to CSR, with its emphasis on supplier codes and compliance, has been marginal to the issue of addressing the position of primary producers. Imposing pro-poor and inclusive procurement on suppliers, with the usual tools of supplier standards and compliance, will not bring about more inclusive markets. Nevertheless, there is a valid debate in rural development about the relative contribution of smallholder production versus plantation wage labour to the rural economy and to poverty reduction. Many wage labourers are the poorest of the poor, what OECD-DAC refer to as Rural World 4 and Rural World 5 (OECD, 2006). The approaches to CSR and business and development have themselves diverged, with 'ethical' being focused on compliance approaches to labour standards, while 'fair' focuses on small-scale producers.

Another debate is between the importance of 'modern' and 'traditional' markets. Many of the models developed around modern restructured chains apply equally well to local and 'traditional' markets. Some emerging modern markets are extremely small, niche and donor-influenced, and a distraction from the priorities of broad-based rural development. Furthermore, even very progressive modern procurement systems can be exclusionary. Producers, intermediaries, buyers and support agencies must evaluate their options very carefully; there may be better rewards in the traditional markets, thanks to the high volume and less stringent standards. We need to know more about the applicability of new business models to trade with traditional markets.

This leads to the importance of acknowledging the risks of the 'new business model' concept. It is an open question whether new business models will benefit the poorest, and, if they do, whether they will ever be sustainable. Over-reliance on markets coupled with voluntary pro-poor initiatives by business misses the point of market governance, whereby genuinely effective business models work best in a strongly supportive policy environment for both producers who want to connect to those chains and those who cannot. Policy attention will always be required to prevent persistent and engrained abuse of power in asymmetric power relations. Business models do not resolve other key issues such as infrastructure investments that may be critical to upgrade excluded producers. These 'hidden costs of inclusion' are not well accounted for in business models, but failure to include them limits the effects of even the most progressive approaches on the rural economy.

References

Abramovay, R. and Magalhães, R. (2007). *Innovative Practice Brazil: access of family farmers to biodiesel markets – partnerships between large companies and social movements.* Regoverning Markets Innovative Practice series, IIED, London. Available at: www.regoverningmarkets.org.

Agropyme (2006). Innovaciones Organizacionales de Pequeños Productores de Vegetales para Participar en Canales de Comercialización Dinámicos en Honduras. Report prepared for the Regoverning Markets Consortium. Agropyme, Agencia Suiza para el Desarrollo y la Cooperación (COSUDE) and Swisscontact, Tegucigalpa, Honduras.

Aramyan, L., Ondersteijn, C., van Kooten, O. and Oude Lansink, A. (2006). Peformance indicators in agri-food production chains. In C.J.M. Ondersteijn, J.H.M. Wijnands, R.B.M. Huirne and O. van Kooten (eds) *Quantifying the agri-food supply chain*, pp. 47–64. Springer, Wageningen, The Netherlands.

Bachev, H. and Manolov, I. (2007). *Bulgaria: inclusion of small-scale dairy farms in supply chains: a case from Plovdiv region.* Regoverning Markets Innovative Practice series, IIED, London. Available at: www. regoverningmarkets.org.

Bell, D.E., Milder, B. and Shelman, M. (2007a). *Vegpro Group: growing in harmony.* Harvard Business School Case Study, December 2007. Boston, USA.

Bell, D.E., Sanghavi, N., Fuller, V. and Shelman, M. (2007b). *Hariyali Kisaan Bazaar: a rural business initiative.* Harvard Business School Case Study, November 2007. Boston, USA.

Berdegué, J. (2001). Cooperating to compete – associative peasant business firms in Chile. Ph.D. thesis, Wageningen University and Research Centre, Wageningen.

Berdegué, J.A., Biénabe, E. and Peppelenbos, L. (2008a). Innovative practice in connecting small-scale producers with dynamic markets. Available at: www.regov erningmarkets.org.

Berdegué, J.A., Reardon, T., Hernández, R. and Ortega, J. (2008b). Modern market channels and strawberry farmers in Michoacán, Mexico: Micro study report. Regoverning Markets Programme. Available at: http://www.regoverningmarkets.org.

Bienabe, E. and Vermeulen, H. (2007). *New trends in supermarket procurement systems in South Africa: the case of local procurement schemes from small-scale farmers by rural-based retail chain stores.* Regoverning Markets Innovative Practice series, IIED, London.

Bouma, J. (2005). Value Chain Workshop: background, methods and management principles. China–Canada Agricultural Development Program. First Training session of Changing Agri-Food Markets, Canada, 9–30 July. Available at: http://www.ccag.com.cn/downloads/training/train_agrifood_market/1st/eng/%5Beng%5DValue_Chain_Workshop.ppt.

Brom, J, 2007. *Best Commercial Practice Code (2000–2006) as an efficient policy innovation to prevent conflict and solve controversies between suppliers, processors and supermarkets.* Regoverning Markets Innovative Policy Series, IIED, London.

Cadilhon, J. (2006). Trading partners self-developing market linkages: Mr Van's let-

tuce supply chain to HCM City (Viet Nam). Presentation to FAO/VECO Workshop on enhancing capacities of NGOs and farmer groups to link farmers to markets, Bali, Indonesia, 9–12 May 2006. Available at: http://www.fao.org/ag/ags/subjects/en/agmarket/linkages/Bali/Cadilhon.pdf.

Camacho, P., Marlin, C. and Zambrano, C. (2007). *Elementos orientadores para la gestión de empresas asociativas rurales.* Plataforma RURALTER, Mesa de trabajo en Desarrollo Económico, Quito, Ecuador, March, 94 pp.

CIAT and CIPASLA (2006). Proyecto Mejorando los Servicios locales de apoyo en Cauca (Colombia) y Yorito (Honduras): Informe del Estudio de la Fase I. Project document presented to NZAID, 32 pp.

CIPS and Traidcraft (2008). *Taking the lead: a guide to more responsible procurement practices.* Chartered Institute of Purchasing and Supply and Traidcraft. Available at: www.traidcraft.co.uk/international_devel opment/policy_work/purchasing_practices.

Concepcion, S.B., Digal, L. and Uy, J. (2006). *Keys to inclusion of small farmers in the dynamic vegetable market: the case of NorminVeggies in the Philippines.* Regoverning Markets Innovative Practice Series, IIED, London. Available at: www. regoverningmarkets.org.

Davies, R. (2007). *The trouble with supermarkets – a way forward.* 2 March 2007. Available at: http://www.seeingthepossibil ities.com/?p = 33.

Eaton, C. and Shepherd, A.W. (2001). Contract farming. Partnerships for growth. FAO Agricultural Services Bulletin 145. FAO, Rome.

Espinal, C.F., Martínez Covaleda, H.J. and Peña Marín, Y. (2005). *La Industria Procesadora de Frutas y Hortalizas en Colombia.* Ministerio de Agricultura y Desarrollo Rural, Observatorio de Agrocadenas, Bogotá, October, 52 pp.

FAO (2008). Contract Farming Resource Centre. Available at: http://www.fao.org/ag/ags/contract-farming.

FAO and CIFOR (2002). Towards equitable partnerships between corporate and smallholder partners: relating partnerships to social, economic and environmental indicators. Available at: http://www.fao.org/DOCREP/005/Y4803E/y4803e00.htm#TopOfPage.

Garside, B., Vorley, B. and MacGregor, J. (2007). Miles better? How 'fair miles' stack up in the sustainable supermarket. Sustainable Development Opinion Paper, IIED. Available at: http://www.iied.org/pubs/display.php?o=17024IIED&n=5&l=45&s=SDO.

Gupta, R. (2008). New models and innovation in India: the Hariyali experience. Presentation to the conference Inclusive Business in Agrifood Markets: Evidence and Action, Beijing, 5–6 March 2008. Available at: www.regoverningmarkets.org/en/file manager/active?fid=828.

Hart, S.L. (2007) *Capitalism at the crossroads: the unlimited business opportunities in solving the world's most difficult problems.* Wharton School Publishing, Upper Saddle River, NJ.

Hayami, Y. and Otsuka, K. (1993). *The economics of contract choice: an agrarian perspective.* Clarendon Press, Oxford.

Hellin, J., Lundy, M. and Meijer, M. (2007). Farmer organization, collective action and market access in Meso-America. CAPRi Working Paper No. 67, CGIAR Systemwide Program on Collective Action and Property Rights, Washington, DC, October.

Hobbs, J.E., Cooney, A. and Fulton, M. (2000). *Value chains in the agri-food sector: What are they? How do they work? Are they for me?* Specialized Livestock Market Research Group, College of Agriculture, Department of Agricultural Economics, University of Saskatechewan, Saskatoon, Canada.

Hu, D and Xia, D (2007). *China: case studies of Carrefour's quality lines.* Regoverning Markets Innovative Practice Series, IIED, London. Available at: www.regoverning markets.org.

Lemeilleur, S. and Tozanli, S. (2006). A win–win relationship between producers' unions and supermarket chains in Turkish fresh fruits and vegetables sector. Paper pre-sented at the International Seminar USAID Regional Consultation on linking farmers to Markets. Cairo, Egypt, 28 January–3 February 2006. Available at: http://eumed-agpol.iamm.fr/html/publications/part-ners/lemeilleurtozanli.pdf.

Lundy, M. (2007). New forms of collective action by small-scale growers. Input for the World Development Report 2008. Rimisp, Santiago, Chile. Available at: http://www.rimisp.org/getdoc.php?docid=9855.

Lundy, M. and Fujisaka, S. (2008). Prototipo de un ecosistema de negocios: Evaluación de la línea de productos SPA de Mabeli S.A. en Totonicapán. Informe Final. Sustainability Insitute, Comisión Presidencial para el Desarrollo Local, CDRO, PNUD y PRONACOM, Ciudad de Guatemala, mayo, 62 pp.

Lundy, M., Gottret, M.V., Best, R. and Ferris, R.S.B. (2006). *A guide to evaluating and strengthening rural business development services. Field Manual.* Centre Internacional de Agricultura Tropical (CIAT), Cali, Colombia.

Mafuru, J.M., Babu, A.K. and. Matutu, T.F. (2007). *Tanzania: impact of market links on horticultural production in the Mara region.* Regoverning Markets Innovative Policy Series. Available at: www.regovern ingmarkets.org.

Medina, R. and Aguirre, M. (2007). *Strategy for the inclusion of small and medium-sized avocado producers in dynamic markets as a result of phytosanitary legal controls for fruit transport in Michoacán, Mexico.* Regoverning Markets Innovative Policy Series, IIED, London.

Ochoa, L. and Lundy, M. (2001). El caso de producción de pasta de ají para export-ación en el Valle del Cauca, Colombia. Prepared for FAO Latin American Regional Office as part of the Estrategia de Alianzas para el Desarrollo de Cadenas Productivas Project. October 2001. Available at: http://www.fao.org/Regional/LAmerica/prior/desrural/agroindustria/pdf/pastaji.pdf.

OECD (2006). *Promoting pro-poor growth: agriculture*. Available at: www.oecd.org/dac/poverty.

Osterwalder, A. (2006). Business Model Template-brief outline of business models. November 2006. Available at http://business–model–design.blogspot.com/2006/11/business-model-template-designing-your.html.

Prahalad, C.K. and Hart, S.L (2002). The fortune at the bottom of the pyramid. *Strategy + Business* 26: 54–67.

Reardon, T. and Huang, J. (2008). Patterns in and determinants and effects of farmers' marketing strategies in developing countries. Synthesis Report: Micro Study. Available at: www.regoverningmarkets.org

Samaratunga, P.A. (2007). *Sri Lanka: innovative practice in integrating small farmers into dynamic supply chains: a case study of MA's Tropical Food Company*. Regoverning Markets Innovative Practice Series, IIED, London. Available at: www.regoverningmarkets.org.

Sandredo (2006). The dynamics of supermarket supplier. Presentation to FAO – Vredeseilanden Sub Regional Seminar on Enhancing Capacity of NGOs and Farmers Groups in Linking Farmers to Markets, Bali, 9–12 May 2006. Available at: http://www.fao.org/ag/Ags/subjects/en/agmarket/linkages/Bali/bimandiri.pdf.

Sharma, V.P. (2007). India's agrarian crisis and smallholder producers' participation in new farm supply chain initiatives: a case study of contract farming. IIM Working Paper No. 2007-08-01, Indian Institute of Management, Ahmadabad. Available at http://www.iimahd.ernet.in/publications/public/FullText.jsp?wp_no=2007-08-01.

Shepherd, A.W. (2007). Approaches to linking producers to markets. FAO Rural Infrastructure and Agro-Industries Division, Rome. Available at: http://www.fao.org/ag/ags/subjects/en/agmarket/linkages/agsf13.pdf.

Shepherd, A.W. and Cadilhon, J.-J. (2009). Commodity associations and their potential role in supply chain development. In: Batt, P.J. (ed.) *Improving the Performance*

of Supply Chains in the Transitional Economies. Proceedings of the UP Mindanao Centennial Conference. Davao, 2008, Banwa (in press).

Shudon, Z. (2008). *China: an example of an agricultural brokers' association: the Tongzhou Agricultural Broker Association*. Nanjing Agricultural University. Regoverning Markets Innovative Practice Series, IIED, London. Available at: www.regoverningmarkets.org

Sporleder, T., Jackson, C. and Bolling. D. (2005). Transitioning from transaction-based markets to alliance-based supply chains: implications for firms. *Choices* 20(4), 275–280.

Stringfellow, R., Coulter, J., Lucey, T., McKone, C. and Hussain, A. (1997). Improving the access of smallholders to agricultural services in sub-Saharan Africa: farmer cooperation and the role of the donor community. ODI Natural Resource Perspectives, Number 20, June 1997.

Tallontire, A. and Vorley, B. (2005). *Achieving fairness in trading between supermarkets and their agrifood supply chains*. UK Food Group, London. Available at: www.ukfg.org.uk

Van der Vorst, J.G.A.J. (2006). Performance measurement in agri-food supply-chain: an overview. In C.J.M. Ondersteijn, J.H.M. Wijnands, R.B.M. Huirne and O. van Kooten (eds) *Quantifying the agri-food supply chain*, pp. 13–24. Springer, Wageningen, The Netherlands.

Vermeulen, S., Woodhill, J., Proctor, F. and Delnoye, R. (2008). *Chain-wide learning for inclusive agrifood market development*. IIED, Wageningen and Wageningen International.

WBCSD and SNV (2008). *Negocios Inclusivos: Iniciativas Empresariales Rentables con Impacto en el Desarrollo*. World Business Council for Sustainable Development (WBCSD) and SNV Netherlands Development Organization. Available at: www.inclusivebusiness.org

Wiboonpongse, A., Sriboonchitta, S. and Khuntonthong, P. (2007) *Innovative management leading housewives' group into* potato supply chain in Thailand. *Regoverning Markets Innovative Practice series*, IIED, London.

World Bank (2006). *Enhancing agricultural innovation: how to go beyond the strengthening of research systems.* World Bank, Washington, DC.

World Bank (2007a). *Horticultural producers and supermarket development in Indonesia.* World Bank, 2007. Report No. 38543-ID. Available at: http://siteresources.worldbank.org/ INTINDONESIA/Resources/Publication/ 280016–1168483675167/Holtikultura_ en.pdf

World Bank (2007b). *World Development Report 2008: agriculture for development.* World Bank, Washington, DC.

Annex 1: Suggested Selected Web Resources.

Empowering Smallholder Farmers in Markets (ESFIM)	www.esfim.org
FAO Linking Farmers to Markets	www.fao.org/ag/Ags/subjects/en/agmarket/ linkages/index.html
FAO Contract Farming Resource Centre	www.fao.org/ag/Ags/contract-farming
Inter-agency web site for the exchange of information on value chains, linkages and service markets	www.bdsknowledge.org
Making Markets Work Better for the Poor project (Vietnam, Cambodia, Laos)	www.markets4poor.org
Regoverning Markets programme	www.regoverningmarkets.org
Sustainable Food Lab	www.sustainablefood.org
	www.iadb.org/csramericas/iaccsr_home.html
Traidcraft – purchasing practices	www.traidcraft.co.uk/international_develop ment/policy_work/purchasing_practices
WBCSD-SNV Alliance on Business Solutions for Development	www.inclusivebusiness.org

7 Corporate Social Responsibility for Agro-industries Development*

CLAUDIA GENIER,[1] MIKE STAMP[2] AND MARC PFITZER[3]

[1]Senior Consultant, FSG Social Impact Advisors, Geneva, Switzerland; [2]Consultant, FSG Social Impact Advisors, Geneva, Switzerland; [3]Managing Director, FSG Social Impact Advisors, Geneva, Switzerland

Introduction

The world can embrace new roles for business in society but achieving progress still requires major and sustained investment from all sectors. This simple fact certainly holds true when considering the agrifood industry's initiatives to improve conditions around food value chains, even if it is at times ignored. At one extreme, companies promote 'responsible' practices that only well-resourced value chain partners can implement; at the other, businesses work hand in hand with other sectors to achieve change that benefits the poor. Similarly, the public sector seems at times unable or unwilling to invest in food value chains as engines of national growth. In other situations it has built on the power of business innovations to reap economic and social benefits beyond the interests of individual companies. This review of corporate social responsibility (CSR) in the agrifood sector provides a landscape of opportunity won and lost and encourages the most promising strategies and collaborative approaches.

There is increased attention to the engagement of business in society. There is growing recognition that private enterprise that operates on an international scale must help find solutions to global problems. Governments, non-governmental organizations (NGOs) and the media have put large companies in the spotlight to account for the social consequences of their activities. As a result, CSR has emerged as an important area of action for large companies globally. For the agrifood sector, which is dependent on natural, human and physical resources, responsible innovation is increasingly being viewed by firms as a corporate and strategic necessity to ensure long-term sustainability.

Few industries have the potential to contribute to development progress on the same scale as the agrifood industry. Its value chains involve millions of

*This chapter was prepared with support from Nestlé and with assistance from the Sustainable Agriculture Initiative (SAI) Platform and the Sustainable Food Laboratories.

people, from farm input providers to consumers, with many from developing countries. A relatively small number of companies have the ability to affect the lives of millions of people and their use of natural resources. Yet the agrifood sector today faces critical challenges: global food demand is due to double in the coming 25 years, requiring an equivalent increase in agricultural production. The growth in demand increases the potential to capture value from agriculture and food production, and could offer large numbers of small-scale producers an opportunity to improve their livelihoods. For this to occur, however, a fair share of the value generated by agrifood chains needs to be captured upstream at the producer level. Particularly in countries where agriculture is a major source of GDP growth, the agrifood industry is of critical importance in fighting poverty and achieving progress towards the Millennium Development Goals (MDGs). The emergence of CSR has played a significant role in stretching the boundaries of action of corporations towards these objectives.

Many companies have engaged in CSR for defensive reasons. While some companies view CSR principally as a public relations tool based on traditional philanthropy, others use it to prevent negative media publicity by imposing 'ethical' codes of conduct within their value chains. Increasingly, too, they cooperate with competitors in the same industry in an effort to set common values, spread risks and shape opinion.

This approach has historical precedence. The knee-jerk reaction of criticism for poor standards has been to 'fix the problem' and to ask business partners in the value chain to do the same. Driven by concerns about food safety, voluntary standards and codes have proliferated into a long list of competing norms, which progressively included a broader set of environmental and social concerns. On the positive side, these norms represent a new frontier of practices leading to more sustainable agrifood value chains;[1] their impact on development objectives, however, is less clear. Standards and codes help achieve progress when they are adopted; yet adoption requires considerable and tailored investment in farmers' aspirations, skills, capital and context. Well-resourced value chain partners (in richer countries) can more easily adopt new practices without external support. Poor rural communities, on the other hand, may find it significantly more challenging to join these 'responsible' supply chains without assistance.

A growing number of leading companies, of which some are featured in this chapter, take a broader view of CSR. They see an opportunity to create 'shared value' through their activities and investments, benefiting all stakeholders in the agrifood value chain including (but not limited to) themselves, as well as the society in which they operate. This motivates such companies to foster stronger ties with local producers, suppliers and communities. Approached in this more strategic and proactive way, CSR can become part of a company's long-term competitive advantage by creating more favourable conditions for business. Such 'integrated' strategies combine business reengineering and strategic social investments: the first, to change how business is conducted; and the second, to improve the context in which such innovations are delivered.

---

[1] For a more detailed description of corporate innovations in socially responsible behaviour as a moving frontier, see Martin (2002).

Looking at such 'proactive' examples from the agrifood sector, we find that corporate initiatives that go deep into rural areas with appropriate resources to transform food value chains tend to present more tangible economic and social benefits than those based on a more defensive view of CSR. Direct investment in communities, as seen through specific case studies and value chain innovations featured in this chapter, enables companies to reconcile their supply chain priorities with community development and environmental gains. These examples are compelling, but alone will not add up to the development of competitive and sustainable national agrifood sectors: projects typically reach select communities for specific objectives of interest to sponsoring companies. What is left are ingredients for scale-up, which must be nurtured by the public sector and civil society as promoters of implementation at national levels.

This chapter examines the principal dynamics created by the defensive and proactive strands of CSR in the agrifood sector.

Approach

This chapter explores the CSR initiatives of agrifood companies that have potential to:

* preserve and improve the natural environment and community welfare; and
* increase the inclusiveness and competitiveness of agrifood companies.

As primary research, the chapter builds on an analysis of a number of standards and codes, as well as several case studies of value chain innovations driven by companies for business and/or philanthropic reasons. It further integrates secondary research from relevant publications on sustainable agriculture.

To understand the current state of standards and codes as tools to promote sustainable agriculture, and the impact they have had, an extensive literature review was conducted and 14 schemes selected for detailed analysis from a list of over 100 (see Annex 1). In addition, ideas and hypotheses were tested with external experts from the agrifood industry, standards organizations and the non-profit sector.

The review of over 40 projects conducted by corporate members of the Sustainable Agriculture Initiative (SAI) Platform was followed by in-depth research into seven case studies, which increased our understanding of value chain innovations in the agrifood industry.

In addition, nine experts from five multinational companies were interviewed by telephone. Secondary research included a review of relevant reports, studies and articles from various sources.

The research examines the two prevailing modes of action at the forefront of agrifood companies' CSR agendas: the roll-out of standards and codes, and value chain innovations. It explores the motivations for implementing such initiatives, and their effectiveness. The chapter further explores the role that government and civil society can play to make CSR a more effective tool in the development of agrifood chains that are equitable and inclusive, as well as competitive.

Standards and Codes: Use of CSR for Risk Management

Agrifood companies use standards and codes as tools to promote sustainable development through their supply chains by influencing suppliers to adopt more environmentally and socially responsible practices. All standards consist of a series of criteria, or rules, with which third-party suppliers are asked to comply (though the number, content and stringency of these criteria can vary substantially between schemes). In many cases, they represent an attempt to bridge the gap between legal and social norms in producer and consumer countries and, particularly on social issues, they may be developed in response to perceived weaknesses in laws and law enforcement.

Companies are motivated to roll out standards and codes for several reasons. First, standards and codes increase the control they can exercise over their supply chains in a cost-effective way (at least for the company), in order to manage risks relating to food safety and corporate reputation. In an industry with such a high degree of exposure to consumers, this control is very important: a company whose food products are contaminated, or whose suppliers exploit child labour, can suffer serious reputational damage. Moreover, legislation passed in Europe, Japan and the USA in the early 1990s requires retailers to conduct 'due diligence' on food safety; companies that cannot demonstrate that their products are safe to eat may not be allowed to sell to these markets at all.

Standards and codes are also important to validate claims of sustainability made for marketing purposes – particularly when they are associated with a label or a charter mark. This has been historically important for niche schemes, such as Fairtrade or Organic produce, which must justify higher prices to consumers. Today, companies are increasingly incorporating sustainability claims into mainstream brand strategies. One noteworthy example is Unilever's decision to ensure all of its Lipton and PG Tips brand tea is certified by the Rainforest Alliance by 2015 (see case study 6 in Annex 2).

For companies that have decided to take action on sustainable development through their supply chains, there are a number of pragmatic, operational reasons for choosing standards and codes over other modes of action. Most important among these is that the approach is well understood. The earliest schemes relating to food safety have been in existence for more than 2 decades. During that time, managers have had ample opportunity to build up experience on how they work and how to implement them on a large scale. As more companies have taken specific action on sustainability, the pool of knowledge in this area has grown further. Managers can access information and support from colleagues and competitors (e.g. through industry platforms such as SAI), as well as from specialized firms such as consultants and IT providers. Schemes also provide companies with a ready-made 'to-do' list of good practices, allowing them to quickly and easily identify the specific changes they hope to induce.

Companies typically adopt two approaches to implement standards and codes. In some cases – Fairtrade, for example – producers are offered an incentive scheme such as a price premium or guaranteed purchasing agreement if they meet the criteria, or work towards doing so. In others, a market entry 'stick' is applied; producers that cannot or do not comply within a given time period are excluded. These approaches may be combined, depending on both the schemes themselves and the companies that use them. There is some evidence that, particularly in mainstream/mass market supply chains, the 'stick' approach predominates (Rotherham, 2005).

The current situation

In recent years, standards and codes have proliferated. Over 100 different schemes covering various aspects of sustainable development can be found by carrying out a simple web search. Their origin explains this complex landscape.

Food safety considerations were amplified in the late 1980s and early 1990s when, after a series of health scares, regulation was passed in European countries, Japan and the USA requiring retailers and, by extension, manufacturers to conduct 'due diligence' against contamination. A parallel movement saw specialist schemes focused on sustainable development emerging under the leadership of NGOs and social enterprises such as the Utz Kapeh Foundation. In recent years, the worlds of food safety and sustainable development have moved closer together. The rise of consumer and activist pressure for more sustainable food production has led many schemes that initially focused on quality and food safety to expand to other sustainability-related issues. Companies saw opportunities to use sustainability as a marketing attribute; illustrated, for example, by Nestlé's Fairtrade-labelled 'Partner's Blend' coffee, or Anheuser-Busch's entry into the organic beer market.

Many companies have developed their own proprietary schemes, aimed at their suppliers. More recently, thinking has evolved from seeking to 'take on the world alone' to a recognition of the need for cross-industry collaboration. There is some evidence that companies are moving away from an in-house approach to adopt independent, third-party frameworks. In 2003, for example, Chiquita applied a proprietary code of conduct; today 100% of Chiquita's owned plantations are certified compliant with GLOBALGAP, the Rainforest Alliance and SA8000.[2]

Partly in response to these trends, the number of independent, third-party standards and codes is increasing. Many are designed specifically for the agri-food industry, and have been developed and backed by governments, NGOs,

---

[2] Editors' note: As we indicated in Chapter 2 (this volume) GLOBALGAP is a private sector body that sets voluntary standards for the certification of agricultural products internationally. The Rainforest Alliance sets standards for environmental and social community sustainability. SA8000 refers to Social Accountability International's system for managing ethical workplace conditions in global supply chains. The Ethical Trade Initiative identifies and promotes good practices in the implementation of codes of conduct regarding labour standards. ISO 14002 is one of the sets of environmental management systems specified by the International Organization for Standardization (ISO).

industry consortia or private organizations. The basic business model of many of these schemes is to build legitimacy and revenue through widespread adoption. Privately owned schemes rely on income from royalties, consultancy fees or franchising rights, while NGOs and government-backed frameworks need to demonstrate certain levels of uptake in order to continue to receive funding. The more general standards and codes, such as those of the Ethical Trading Initiative, SA8000 and ISO 14002, do not rely only on the agrifood industry for survival.

Detailed analysis of 14 schemes

Fourteen independent, third-party standards and codes have been analysed in detail, yielding a number of insights. While each covered a slightly different set of issues, it was possible to map the main areas that schemes aim to address, across four basic categories:

- **Environmental** criteria focus on:

    - ecosystems and biodiversity (e.g. provisions to protect virgin forest);
    - natural resource inputs (e.g. water use, soil quality);
    - man-made inputs (e.g. agrochemicals, pest control, GMOs);
    - energy use and GHG emissions;
    - waste management;
    - production practices (e.g. crop rotations, site selection, animal welfare, overfishing).

- Criteria on **labour conditions** can be grouped into:

    - occupational health and safety;
    - terms of employment (e.g. pay, hours, contracts, regularity of work);
    - human rights in the workplace (e.g. right of association, rights for casual workers, no forced or child labour, non-discrimination);
    - general employee and family welfare (e.g. housing, access to education and healthcare);
    - energy use and GHG emissions;
    - waste management;
    - production practices (e.g. crop rotations, site selection, animal welfare, overfishing).

- Criteria relating to the **benefits to the local economy/community** include:

    - producers' economic viability;
    - flow of economic benefits to workers and the local economy;
    - social and economic rights of others (e.g. indigenous land rights, local consultation);
    - business ethics (e.g. fair dealing, no corruption, market transparency);
    - education and role modelling (e.g. open days).

Table 1. Comparison of issue coverage of 14 independent standards and codes.

	Environment						Labour conditions				Local economic/ community benefits					Food safety and quality			
	Ecosystems & biodiversity	Natural resource inputs	Man-made inputs	Energy use and GHG emissions	Waste management	Production practices	Occupational health & safety	Terms of employment	Human rights in the workplace	General employee/family welfare	Producers' economic viability	Flow of economic benefits	Social/economic rights of others	Business ethics	Education & role-modelling	Traceability	Hygienic production & handling	Quality of inputs	Quality of management systems
Utz Certified	●	●	●	●	●	●	●	●	●	●						●	●	●	●
EISA	●				●	●	●	●		●		●			●		●		
SAI Principles & Practices for Sustainable Production (Cereals)	●	●	●	●	●	●	●	●	●	●	●	●				●	●	●	
IDF/FAO Guide to Good Dairy Farming Practice					●	●											●	●	
Fairtrade Standards	●	●	●	●	●	●	●	●	●	●		●							●
SA8000							●	●	●	●									
Roundtable on Sustainable Palm oil	●	●			●	●	●	●	●		●		●	●	●				
Basel criteria for Responsible Soy Production	●	●	●	●	●	●	●	●	●		●	●	●			●	●	●	
Marine Stewardship Council	●				●	●							●						
Common Code for the Coffee Community	●	●	●	●			●	●	●	●			●	●		●			●
Ethical Trading Initiative							●	●	●	●									
SCS-001	●	●	●	●	●	●	●	●	●	●	●	●	●			●	●		●
Rainforest Alliance/SAN	●	●	●		●	●	●	●	●	●		●	●		●	●			
GLOBALGAP	●	●	●	●	●	●	●									●	●	●	●

- Lastly, sustainable agriculture standards and codes may set down requirements on **food safety and quality**, in particular:

  - traceability;
  - hygienic production and handling;
  - quality of inputs (seeds, feeds, etc.);
  - quality management systems.

In addition to criteria based on some combination of these categories, all of the standards and codes studied include requirements relating to general management issues, such as record keeping or planning, and to compliance with the scheme itself.

Table 1 displays how each covers the respective issues.[3]

---

[3] Note that simply covering the same issues does not necessarily mean that standards are equivalent: see Annex 1 for further discussion.

Equally interesting are the areas that standards and codes do not seem to address. None addressed gender issues and, aside from a footnote in the introduction to SCS-001 expressing a desire to incorporate nutrient density into a future version of the standard, none addressed the issue of food security. Additionally, only four schemes made any provision for the economic viability of producers themselves, though this is also arguably implicit in the structure of the Fairtrade scheme.

Where standards and codes do not seem to explicitly differentiate is around contextual conditions. The environmental effects of particular practices can vary dramatically depending on the weather, time of year, soil type, level of water depletion within a river basin and many other factors; yet no schemes could be found that took account of such differences, even when specifying particular processes. Similarly, what makes sense for large, well-resourced farms may be meaningless for smallholders. Some schemes, such as the Ethical Trading Initiative, Fairtrade and the Basel Criteria, have begun to develop altered or slimmed-down versions aimed at smaller producers, but this is not yet the norm.

It is also important to bear in mind that the fact that a particular scheme incorporates criteria on a wider range of issues does not necessarily mean it is better. Indeed, one major concern that became clear during the research for this chapter was the lack of evidence as to whether particular standards and codes are effective at all on any given issue.

Challenges going forward

The fact that the agrifood industry is investing so heavily in standards and codes that seek to address sustainability issues is encouraging, and demonstrates an important shift in companies' willingness to engage on such topics. Nevertheless, challenges are faced by all parties in harnessing this activity to drive sustainable development.

Lack of impact assessment

While there is a growing body of literature on the (indirect) economic impact of mainstream standards and codes on producers and rural communities, particularly in developing countries, there is surprisingly little evidence of whether or not they have a significant positive impact on environmental, labour or community development issues.

The handful of studies that consider the direct impact of specific standards and codes on producers, rural communities and the environment suggests that they are more effective on observable issues (i.e. those that do not require subjective or culturally specific judgements to determine compliance or progress) that do not vary much according to context, though there is too little information to fully validate this hypothesis. For example, the 2006 impact assessment of the Ethical Trading Initiative found that application of that code had had a major impact on workers' health and safety (Barrientos and Smith, 2006). This topic primarily concerns tangible issues such as the existence of procedures and officers, safe use of chemicals and the quality of drinking water, all of which can be directly observed and are fairly universal in scope. By contrast, almost no change was found in, for example, freedom of association, reflecting the fact that such matters are very difficult to assess and are highly dependent on contextual factors like cultural norms and forms of local and national governments. A study of Zambian exporters' views of GLOBALGAP, conducted by the Natural Resources Institute, points to a similar conclusion: improvements were particularly cited on workers' health and safety, traceability and food safety (Graffham and Vorley, 2005).

Current evidence suggests that the question of impact assessment has not yet been satisfactorily resolved. The SAI Platform has started work on developing a cost–benefit tool to assess different schemes, and a number of papers, including, notably, one from the ISEAL Alliance, have called for more systematic impact assessment. As yet, though, such efforts seem still to be at an early stage of development.

In many instances, therefore, it is not clear whether standards and codes are an effective tool for promoting sustainable development at all (though, of course, it is important to remember that there is also insufficient evidence to declare them ineffective). This lack of impact data means that schemes are evolving and competing with each other in a vacuum: companies select which schemes to implement (and governments and civil society select which ones to endorse) based on attributes such as reach, credibility and business friendliness, but not effectiveness. As a result, scheme developers have little incentive to calibrate and focus their criteria to provide the best return on investment, or to adapt their approach in light of new research.

Lack of mutual recognition

In seeking to reach scale, independent standards and codes are increasingly in direct competition with each other, and efforts to rationalize and harmonize the landscape through mutual recognition of schemes have been limited. There have been some attempts to establish equivalence between non-proprietary schemes, such as GLOBALGAP's initiative to compare itself with government-backed national standards. Governments and international organic associations[4] have made some progress in harmonizing the various national organic schemes. However, for the most part, it is probable that this activity is driven more by a desire to establish dominance than to rationalize the field. There are few examples of independent third-party schemes benchmarking themselves against each other.

As a result, producers and consumers are faced with a large number of different labels and options for certification, all of which claim to do a similar job, but with slightly different emphases. The resulting confusion can lead to consumer cynicism or indifference, reducing companies' opportunity to claw back their investment through higher prices (and hence the incentive to invest

---

[4] The International Organic Accreditation Service (IOAS); the International Federation of Organic Agriculture Movements (IFOAM).

in the first place). Furthermore, producers that supply more than one company may be pulled in different directions, magnifying the perceived risks to the producer of 'backing the wrong horse', and making it more likely that they will either follow several methods half-heartedly or do nothing at all. Conversely, there is some evidence (from the ETI impact report) that when all buyers are pulling in the same direction the likelihood of more sustainable practices gaining traction increases (Barrientos and Smith, 2006).

One-size-fits-all solutions

In terms of content, there is a tension between the reach of schemes in terms of products, issues and geographies covered, and the level of detail into which they can go while remaining relevant. Farming practices are highly contextual: soil conditions may change from one side of a field to the other; actions that may save water in one river basin may have no impact in another; and freedom of association may be very important in the context of a large estate, but irrelevant to a family-run smallholding. Yet standards and codes are inherently inflexible and schemes, especially independent ones, need to achieve some degree of scale in order to be credible to consumers and financially viable.

This inflexibility leads to a further concern. The way many standards and codes work, as a kind of pass or fail test of sustainability, is at odds with many of the actual issues encompassed, which tend to change over time, and for which 'success' may not be clearly defined or agreed upon. Sustainable development may be better thought about in terms of continuous improvement rather than 'right' and 'wrong' answers. Our understanding of what this means on specific issues, and how it translates into action, can change rapidly as new information comes to light.

Implementation costs for small-scale producers

On the economic side, widespread implementation of standards and codes seems to accelerate vertical coordination in supply chains (i.e. more production is for specific customers rather than for the open market) and the consolidation of producers into fewer, larger entities. Evidence suggests that this restructuring of the rural economy is an integral part of wider economic development, as the best producers continue to farm larger areas more efficiently, while the majority of the population move to other and for them, more lucrative forms of economic activity (World Bank, 2007).

What is striking is the speed at which this transition occurs when standards and codes are applied. Maertens and Swinnen (2007) studied the economic impact of the roll-out of the *origine Sénégal* label on fruit and vegetable producers in Senegal between 2000 and 2005. During this period, the number of small-scale growers with export contracts fell by 72%, and the three largest exporters increased their market share from less than half to 66% of all exports.

Such developments are primarily because implementing standards and codes is very expensive for producers. Asfaw *et al.* (2007) estimate the first year costs for a small Kenyan bean grower to implement EurepGAP (now GLOBALGAP) as part of a consortium to be around 37,000 Kenyan shillings (KSh),[5] of which around 6500 KSh are recurring costs; the same study puts farmers' average gross annual income from export vegetables at 33,864 KSh (Asfaw *et al.*, 2007). Graffham and Vorley (2005) paint an even starker picture: they estimate that without donor support, first year EurepGAP implementation costs can amount to as much as 160% of annual income for smallholders with 0.2–0.6 ha of land. Clearly, large and well-resourced producers are much better able to absorb such costs, and so have a strong advantage when standards and codes are used as a market entry criterion.

Given that the *World Development Report 2008* (World Bank, 2007) describes competitiveness and smallholder participation as 'essential to link agricultural growth to development', the reader would be forgiven for concluding that standards and codes are an economic 'bad' in development terms. However, the picture is more nuanced than this. For example, the same Maertens and Swinnen study concluded that while *origine Sénégal* was bad news for small contract farmers, it in fact represented a net gain for the economy overall as export production increased from 9000 t in 2000 to 16,000 t in 2005. Moreover, while smallholders lost out, the very poorest rural households, without access to sufficient land or skills to participate in the export market, may have gained by being able to find employment on large estates. In addition, some of the smallholders survived by diversifying into other activities.

There is also some evidence that this acceleration of economic restructuring may be due to the way standards and codes are applied, as opposed to anything inherent in them as a tool. A recent study of Lecofruit, a fruit and vegetable exporter in Madagascar that buys high-quality, certified beans and peas from some 10,000 smallholders (each farming around 1 ha), found significant benefits for growers, including improved productivity (not just for export crops, but for staples such as rice), greater income stability and shorter lean periods (Minten *et al.*, 2005). However, Lecofruit did not simply apply a standard in isolation as a purchasing criterion, but included it in a deliberate policy to engage with smallholders that also encompassed long-term micro-contracts, access to 300 agricultural extension workers, information on composting procedures and loans of seeds and fertilizers to be repaid in kind. The Lecofruit example shows that it is possible to implement a standard or code without excluding small-scale producers, but only as just one part of a wider, ongoing package of intensive support.

Even where there is a clear willingness to implement a specific scheme, suppliers may lack access to the infrastructure and resources needed to do so, particularly in developing countries. Standards and codes require significant investments in supporting infrastructure: reliable and well-regarded inspection and accreditation capacity is needed; extensive training is required, including poorly educated or illiterate producers; materials such as specialist clothing or IT equipment may not be available locally. As we have seen, implementing a scheme can be expensive for producers, too, and this risks driving cash-strapped farmers out of business entirely. Simply subsidizing this cost may not be the

[5] US$1 = 66.9 KSh on 1 November 2007.

answer: evidence from Kenya suggests that donor financing of implementation has distorted the market for inspection and accreditation services, with local suppliers charging up to four times as much as their European equivalents (Graffham and Vorley, 2005).

Value Chain Innovations

Although a one-size-fits-all approach to sustainable development based on the wide application of a particular standard or code may indeed work for companies that have well-resourced suppliers, the more visionary agrifood companies are also starting to realize its limitations. They are adopting innovative approaches to CSR that take a more holistic view of the agrifood value chain and are tailored to comprehensively address specific key issues and situations. Most of these initiatives are designed in multi-stakeholder partnerships (including private and public sector players) to meet common objectives, and are rarely imposed by companies as the leading player. Such initiatives usually start as pilot projects, with the aim of scaling up those that are successful.

Rationale, approach and challenges

Corporate value chain innovations are not a new phenomenon. Nestlé, for example, has been active in India since the early 1960s in the Moga district, not just by opening a dairy factory, but also by encouraging the systematic development of milk production, injecting knowledge and technologies into the system.[6] The company's effort had a multiplier effect on the development of the region. Unilever, too, started an SAI based on economic, social and environmental dimensions roughly 50 years ago. What is new in recent years is the multiplication of such transformative projects, their level of ambition and their positioning within broader CSR activities.

The rationale for investing in value chain innovations can have multiple origins. The company may be trying to stay competitive by emphasizing quality, safety, traceability to small-scale farm operations, local processing, production and distribution and measurable environmental and social benefits. The business case for such innovations is typically aimed at securing the availability of raw materials while improving a community's environment, capturing value through vertical integration and sharing it more equitably with producers.

Two main approaches to value chain innovations can be observed: a bilateral buyer or producer initiative and a multi-stakeholder partnership model involving e.g. farmers' associations, small and medium enterprises (SMEs), the public sector and civil society (a majority of cases). Typically, the more multifaceted the ambitions, the more stakeholders tend to be directly involved in the initiatives. Participation can entail knowledge sharing, endorsement and promotion, implementation or scale-up.

---

[6] Collection centres were set up, farmers were trained, clean drinking water was provided to schools and many more activities took place.

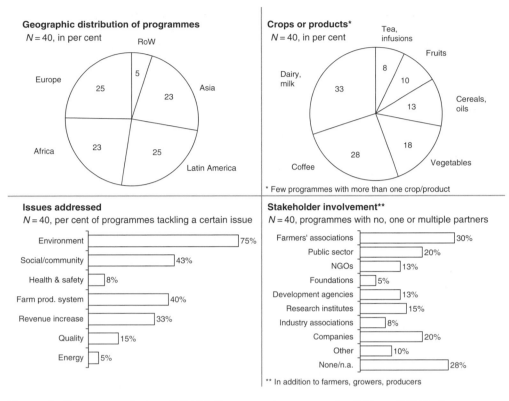

Figure 1. Programme focus of SAI Platform member companies.[7] (From SAI Platform Mapping, 2007. Mapping by FSG Social Impact Advisors, 2007.)

Pilot projects tend to concentrate on one crop, one farm production system or one supply chain, and strive for a combination of technical, financial and educational transfer, as well as capacity building. Figure 1 shows the focus of value chain innovations undertaken by member companies of the SAI Platform, a food industry body supporting the development of sustainable agriculture.[8] Initiatives are distributed fairly equally between Europe, Asia, Africa and Latin America. Half of the projects target the dairy and coffee industry, while the rest deal with vegetables, fruits, cereals, oils and tea. Seventy-five per cent of all projects cover environmental aspects (water, soil, forest, air, biodiversity, ecology); nearly half include social and community aspects (including gender and child labour), and an equal amount look at the farm production system. One-third of all projects aim at improving revenues of producers. Issues like health and safety, quality and energy are tackled specifically by only a minority of projects.

---

[7] Covering programmes from two-thirds of SAI Platform members (i.e. 14 companies out of 22).
[8] Founded by the global agrifood companies Danone, Nestlé and Unilever, it has 22 members today.

A vast majority of projects are launched in partnership: 30% involve farmer associations, 20% the public sector, nearly 20% development agencies or foundations, nearly 30% NGOs or research institutes and nearly 30% other companies or industry associations.

In the last few years, industry or multi-stakeholder dialogues have encouraged corporate innovation on sustainable agrifood value chains and the exchange of good practices. In addition to the SAI Platform, the Sustainable Food Lab is a multi-stakeholder initiative to bring sustainability to mainstream food systems. Agrifood companies, NGOs and others are working together to find innovative ways of addressing persistent problems; such as poverty, in producer nations. The Stone Barns Sustainable Agri-Philanthropy Initiative (Schumacher *et al.*, 2004) is another forum, which has convened major actors to discuss their philanthropic practices.[9]

While case studies selected to illustrate value chain innovations cannot cover the full range of innovative projects launched worldwide, they nevertheless provide concrete insight into how these proactive CSR initiatives shape up. Table 2 provides an overview of the cases detailed in Annex 2. The cases cover three continents (Africa, Asia and Latin America) and several product segments (dairy, crops, fruits and vegetables, tea), present different multi-stakeholder partnerships initiated by six global agri-business players and various intensities of public sector involvement, and cover the entire agrifood value chain (from pre-production to distribution).

While innovations often do embrace the entire value chain, most are demand-driven (i.e. initiated by manufacturers and retailers[10]) and focus on producers, SMEs in the supply chain and economic development at community level. Figure 2 provides a schematic view of supply chain players and illustrates the value chain coverage of each case study.

Before discussing the nature of impact and effectiveness achieved by the specific case studies (Annex 2), it is worth isolating their innovation component:

- In India, SABMiller and Cargill partnered with a foundation in Rajasthan and received the support of regional authorities to lay the ground for developing a viable barley malt industry in the state.
- In China, Nestlé facilitated an innovative approach for its milk suppliers in Shuangcheng, in north-eastern China during the period 2004–2007, to convert the manure from dairy cattle into biogas, to be used for cooking, heating and electricity generation. From the very beginning, the initiative was endorsed and strongly supported by the local authorities, and received further support from the central Chinese government. Chinese authorities are now replicating and scaling up the approach nationwide and extending it to pig farms.

---

[9] It was shown that only a minority of philanthropic giving by agri-businesses was related to their core business, e.g. agriculture systems in developing countries. The vast majority is devoted to community improvements in developed countries.

[10] When asked in surveys, consumers indicate caring about traceability and sustainability; however, the buying behaviour of a large majority signals that price still dictates their choice.

Table 2. Case profiles.

Company	Product/focus	Purpose	Value chain coverage	Partners	Public sector involvement
Brazil: Syngenta	Water recovery	Philanthropic contribution to improving farmers' environment	Pre-production (initiated by supplier of input factors)	Cooperatives, regional authorities	Moderate (promotion, scale-up)
China: Nestlé	Dairy	Biogas production from manure, reducing environmental impact of farming	Production, facilitated by manufacturer/ distributor	Local authorities, central government	Strong (driving, funding, scale-up)
India: Reliance Retail	Fresh fruits & vegetables	Direct sourcing from farmers; share value	Distribution, trading, retail	Farmers (collection centres)	Minimal (legislative framework, licences)
India: SABMiller/ Cargill	Barley	Secure local high-quality production, improve smallholders income	Production, trading, processor/ manufacturer driven	Morarka Foundation, public sector	Strong (facilitation, agri-research, seeds, licences)
Tanzania: Unilever	Allanblackia	Set up commercial supply chain for little-known product	Production, processing, sales (including finding purchaser)	NGOs, research institutes, development agencies	Moderate (agri-research, tree nurseries, endorsement)
Kenya: Unilever	Tea	Certification of sustainable production, signal to market with oversupply	Production, trading, manufacturing, marketing, retail (demand driven)	Producer association, development agency	Minimal (smallholder support, accompanying measures)
Bangladesh: Grameen Danone	Dairy (yogurt)	Set up supply chain adapted for local conditions (poor segments of society)	Production, transport, manufacturing, distribution	Joint venture between Danone and Grameen Bangladesh	Minimal (licences)

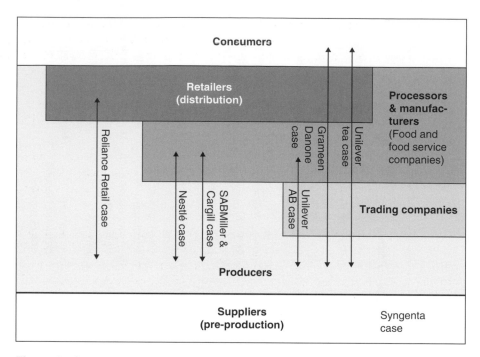

Figure 2. Case study coverage of main supply chain groups. (Adapted from SAI Platform.)

- With its philanthropic project 'Agua Viva', Syngenta has since 2004 contributed to the recovery of over 2100 water heads in Brazil. In collaboration with local cooperatives, Syngenta is providing technical assistance and know-how to farmers to recover their endangered and polluted water heads. As a result, springs are sanitized and protected, the quality of water increases and the water supply is secured for local communities. This project has been a tremendous success locally, winning praise from the Brazilian authorities, as well as both national and international awards.
- Reliance Retail India are directly sourcing fresh agricultural produce from thousands of farmers through collection centres. They are providing a guaranteed market for the farmers' produce, reducing transaction costs and training them in better and more sustainable farming practices. This initiative has resulted in increased income and upgraded skills for the farmers, a reduction in spoilage of produce (by up to 35%), and better quality products for Reliance retail stores. The exclusion from the supply chain of traditional traders, however, led to protests, raising the question of how to evolve potentially obsolete supply chain structures.
- Unilever is working with the Rainforest Alliance to certify tea purchased for its leading Lipton brand by 2015. As the world's largest tea buyer, Unilever intends to send growers a clear signal in favour of sustainable tea production, emphasizing a long-term business model carried by quality rather than by quantity in a market characterized by oversupply.

- The traditional approach to expanding an international business in emerging markets is to target segments likely to find a company's product competitive. Because traditional brands and modes of production are adapted to high-purchasing-power countries, this strategy typically limits developing country sales to the wealthiest customers. Danone Grameen is turning this logic upside down in Bangladesh, rethinking the entire value chain for yogurt production and marketing so that it employs and serves the poorest segments of society.

Evidence of impact can be found in the various initiatives in terms of quality, health and safety improvements, better environmental indicators, higher productivity and economic development, i.e. income growth and diversification, as well as job creation:

- Quality and productivity: in Rajasthan, farmers participating in SABMiller's programme increased significantly their barley yield, by 20% (even if it cannot directly be attributed to the initiative). Over a 2-year period, involved districts achieved an average yield of 33.3%, while the average of Rajasthan state was 13.2%. The quality increased as well, from feed-grade to malt quality. Again in India, Reliance Retail has brought improved quality to its retail stores by collecting fresh fruits and vegetables directly from the producers: spoilage of produce has been reduced by up to 35%.
- Health and safety, environmental indicators: in Brazil, Syngenta has contributed to the health and sustainable water supply of around 2700 families and their livestock by recovering around 2000 endangered or contaminated fountainheads in Brazil since 2004. In China, the biogas project with thousands of generators installed, allowed a reduction in the risk of water contamination by improperly stored manure from dairy farms.
- Income growth, diversification and job creation: in Tanzania, the creation of a supply chain for Allanblackia (AB) provides an additional source of income for farmers. The average AB earning per farmer per year has increased from US$60 to US$140 for 6000 farmers, and should reach US$500 by 2016 for 25,000 farmers. In addition, 45 full-time positions have been created for managing the buying centres. In Bangladesh, the Grameen Danone joint venture predicts that 1000 livestock and distribution jobs will be created. In China, the biogas produced from manure allows dairy farmers to cook three hot meals per day from this new energy source.

Although these value chain innovations clearly show impact, as long as they remain in pilot phase, they are only 'pockets of social progress' that remain minor in comparison with the core business of agrifood industries, or with large-scale initiatives from NGOs and multilateral agencies. To expand their benefit for society, they require scaling up and integration into national agrifood competitiveness policy.

Such a scaling-up process requires considerable resources. Management time is needed to plan the replication, adapt the approach to a larger scale or a different context (e.g. a new country), learn from strengths and weaknesses of the original project, and convince players along the value chain to adopt new practices.

Companies, donors and governments all have a role to play in contributing directly with resources (funds, production factors, capacity building, etc.), or in helping to find other funding sources. Capacity building is often needed at farm level, for logistics, processing and even branding and marketing of produce. Production factors (seeds, plants, etc.) also need to be supplied in sufficient quantities. To scale up a pilot project into a meaningful improvement can require considerable time, even several years in some cases.

In order to be sustainable over time, value chain innovations also need to be grounded in a solid business case, i.e. at some given point (though not immediately) they need to be viable on their own, without additional financial support. If the innovation disappears as soon as project funding ceases, it cannot be considered successful.

This does not mean that value chain innovations are unsustainable as such, but rather that integration of innovation at community, district and national levels needs to be managed and planned carefully for benefits to accrue to an extent where national growth indicators can be affected.

The Role of the Government and Civil Society

CSR initiatives are led and implemented primarily by agrifood companies themselves. None the less, governments and civil society have an important role to play in harnessing the growth in CSR activities for sustainable development of the agrifood sector. They can create the conditions for that growth to continue, both by helping companies to act responsibly and by building an enabling environment.

Multilateral organizations and development agencies are also taking a keen interest in CSR: the UN Global Compact and the Global Reporting Initiative are examples of high-profile public sector initiatives that monitor company performance and publish results globally. Moreover, the emergence of industry platforms and multi-stakeholder bodies, such as the SAI Platform, the Sustainable Food Laboratory and the Ethical Trading Initiative, is one of the most striking recent trends in the agrifood sector and presents a potential new force in the facilitation and documentation of CSR practices.

Standards and codes

Multilateral agencies, governments, industry platforms and civil society need to work together to harmonize standards and codes and make their implementation more accessible and effective. This should be done through measuring and publishing data that will inform impact assessment and help schemes become more locally relevant. Governments can further ensure that good practices are replicated by endorsing those schemes which are most effective. Such endorsements are a key factor in the competition between independent schemes (e.g. Kenya's adoption of GLOBALGAP as a basis for its own national standard). Additionally, industry associations with support from government and NGOs can offer technical assistance to producers, such as training and extension services, and strengthen the local certification and accreditation infrastructure. Finally, governments can play a role in managing the economic effects of standards and codes on small-scale producers.

Value chain innovations

For value chain innovations, governments, local authorities and civil society can smooth the progress and magnify the impact of initiatives in three ways: by creating an enabling environment (adapting legislation, investing in market infrastructure); by participating in pilot projects (facilitating access to production factors such as trial seeds, providing logistical support and playing a facilitator role); and by taking an active role in their scale-up (endorsing replication, co-financing, building capacity and actively participating in industry platforms).

The role taken on by the authorities of Rajasthan offers an example of such active public sector participation, as does the engagement of Chinese authorities in the biogas project facilitated by Nestlé.

The Regoverning Markets Programme[11] is an example of a supportive public sector/civil society initiative. Its aim is to provide strategic advice and guidance to the public sector, agrifood chain actors and civil society organizations including economic organizations of producers on approaches that can anticipate and manage the impacts of the dynamic changes in local and regional food markets. The programme emphasizes analysis of the food industry segments, i.e. retailing, processing and wholesaling.

Smoothing economic transition

All developing and transition economies have seen significant reductions in the number of people employed in agriculture and corresponding increases in productivity. This points to a key long-term role for governments and civil society: helping small farmers to diversify their income sources as well as facilitating a smooth transition out of agriculture. Policy priorities should include creating rural off-farm jobs, increasing accessibility to education and training and providing 'safety nets' to the chronically poor. The example of Unilever's application of a sustainability standard to Lipton tea highlights the need for government in facilitating this transition (see case 6 in Annex 2).

Conclusions and Recommendations

Two types of initiatives have emerged at the forefront of agrifood companies' CSR agendas: standards and codes, and value chain innovations. The initiatives profiled show considerable promise in increasing the sustainability of the

---

[11] www.regoverningmarkets.org.

agrifood sector. However, key challenges in improving their effectiveness remain, such as dealing with the lack of equivalence between different standards and codes, as well as developing more context-specific initiatives that effectively encourage the participation of smallholders in global supply chains.

The increase in corporate activity aimed at benefiting society over the last 2 decades, as evidenced by the rise in CSR, has been accompanied by a keener interest from governments and civil society. In addition, the emergence of industry platforms and multi-stakeholder bodies is one of the most striking recent trends in the agrifood sector, and has the potential to become a powerful force in the facilitation and documentation of CSR practices.

The research has highlighted four concrete areas that require attention from industry, civil society, governments and multilateral agencies:

- First, there is a need to collaborate and invest at the international level to make standards and codes more effective, efficient and accessible through an agenda of impact assessment, mutual recognition and knowledge transfer. As the industry has strongly promoted such standards it would be logical to embed such a process among its champions, such as the SAI Platform and the Sustainable Food Laboratory, while maintaining a close dialogue with civil society, multilateral agencies and other interested stakeholders. These facilities form the hub of leading companies and are connected to crop-specific groups working on standards and codes, as well as on value chain innovations. Funded through corporate membership fees, they do not, however, presently have the resources to orchestrate a comprehensive agenda of standards and codes alignment, and to provide knowledge transfer on a global scale. Recognition and support as hubs of progress might also help assemble a broader membership of companies, including agricultural technology providers, food manufacturers and retailers.

- Second, multiple-stakeholder agro-industry platforms and sector councils at the national level are required to coordinate and monitor implementation of standards, codes and value chain innovations, for mutual support and the sharing of good practices. These might be linked to the international industry platforms mentioned above, mirroring, for example, the structure of the World Business Council for Sustainable Development and its national councils. The role of these national facilities is to facilitate the local integration of value chain innovation into national agrifood competitiveness policies and the work of national agricultural extension services. National platforms need to exist to bring local corporate entities together with national policy makers and civil society with the aim to facilitate adoption of standards and codes for relevant crops and producers at national levels and to proactively identify and support the scale-up of high-potential innovations that help poorer suppliers join higher value supply chains.

- Third, national governments, multilateral organizations and donors should proactively identify and support the scale-up of high-potential innovations, including facilitating the adoption of standards and codes by specific producer groups where this could be beneficial. While CSR innovation in the agrifood sector is being led primarily by agrifood companies, the public sector plays an essential role in facilitating their engagement with producers, creating an enabling environment and expanding lessons from pilot projects to national and international levels.

- Fourth, the evidence points to a key long-term role for governments and civil society in helping diversify small farmers' income sources, including off-farm work, and facilitating their gradual transition out of agriculture into other economic sectors.

References

Asfaw, S., D. Mithöfer and H. Waibel (2007) *What Impact Are EU Supermarket Standards Having on Developing Countries Export of High-Value Horticultural Products? Evidence from Kenya*, Leibniz University of Hanover.

Barrientos, S. and S. Smith (2006) *The ETI Code of Labour Practice: Do Workers Really Benefit?* Institute of Development Studies, University of Sussex.

FSG Impact Advisors (2007) SAI Platform Mapping, unpublished document. Geneva, Switzerland.

Graffham, A. and B. Vorley (2005) *Standards Compliance: Experience of Impact of EU Private & Public Sector Standards on Fresh Produce Growers & Exporters in Sub-Saharan Africa*, NRI/IIED.

ISEAL Alliance (2006) *ISEAL Code of Good Practice for Setting Social and Environmental Standards.*

Maertens, M. and J. Swinnen (2007) *Trade, Standards and Poverty: Evidence from Senegal (revised version)*, Catholic University of Leuven.

Minten, B., L. Randrianarison and J. Swinnen (2005) *Global Retail Chains and Poor Farmers: Evidence from Madagascar*, Catholic University of Leuven.

Rotherham, T. (2005) *The Trade and Environmental Effects of Ecolabels: Assessment and Response*, UNEP.

Schumacher, A., A. Hance, B. Wells, N. Agarwal and N. de Beaufort (2004) *Stone Barns Sustainable Agri-Philanthropy Initiative: Current Status and Projects*, W. K. Kellogg Foundation/Stone Barns Center for Food and Agriculture.

Vorley, B. (2003) *Food, Inc.: Corporate Concentration from Farm to Consumer*, UK Food Group.

World Bank (2007) *World Development Report 2007: Agriculture for Development.*

Useful Websites

- www.agrifoodstandards.net
- www.regoverningmarkets.org
- www.saiplatform.org
- www.sustainablefoodlab.org

Annex 1: Further Comparative Analysis of Standards and Codes

In general, the most important area of differentiation between schemes seems to be the product or product category focused on, such as seafood, dairy, fruits and vegetables. In addition to the product-specific schemes, many of which are among the more recent arrivals on the scene, the majority of proprietary and 'general' sustainable agriculture standards and codes (e.g. SCS-001) include modules that relate to individual products or product categories. There is some variation by geography, too: in some cases this is intentionally designed into a scheme. For example, the EISA-integrated farming framework explicitly targets European farmers. For other examples, geographic reach is limited to areas where the target crop is produced, or where inspection and accreditation infrastructure exists.

For most of the schemes analysed, changes in understanding of sustainable development are reflected through periodic updates. These revisions seem to be led more by expert panels than by wide consultation with producers. While some schemes discuss the need for farms to demonstrate improvement between audits, only three mentioned continuous improvement as a specific criterion or method of incorporating new thinking: the Basel Criteria for Responsible Soy Production; the Roundtable on Sustainable Palm Oil; and the SAI Platform Principles and Practices for the Sustainable Production of Cereals. The primary reliance on expert leadership also seems somewhat at odds with best practice in standard setting identified by the ISEAL Alliance, a group of standard-setting and compliance organizations, which recommends that 'Standard-setting organizations...identify parties that will be directly affected by the standard and proactively seek their contributions' (ISEAL, 2006).

Furthermore, wide variances were found between the stringency and specificity of criteria from different frameworks relating to the same issue. In general, five basic levels of specificity can be identified:

- characterizing what compliance looks like (e.g. 'farm provides a healthy and safe working environment');
- requiring compliance with relevant local laws (e.g. 'local health and safety regulations to be adhered to');
- requiring a policy or plan to be drawn up (e.g. 'farm shall have a health and safety policy');
- suggesting or stipulating a specific process or set of processes be followed (e.g. 'relevant fire safety equipment to be installed in all buildings');
- setting quantitative performance criteria (e.g. 'fewer than two accidents per 100 employees per year').

Which level of specificity is appropriate in which circumstances will vary markedly according to the context in which the scheme is applied and the method which is used to apply it (e.g. it may not be helpful to specify performance criteria on issues that may not be readily measured).

Most of the standards and codes analysed also have different grades of criteria, including a core of baseline provisions that must be complied with in full, and others which may be only partially met, or which only require producers to set out how they intend to work towards meeting them, over a given period.

Annex 2: Selected Cases

Case 1: SABMiller and Cargill improving barley production in India with partners

Since 2005, SABMiller and Cargill have been working together with the Government of Rajasthan, the Morarka Foundation and local farmers to support the development of a viable barley malt industry to supply local breweries.

SABMiller has ten breweries in India. The malt group of Cargill's food division works in partnership with it to grow malting barley. Rajasthan produces around 430,000 t of barley a year. The barley has been of poor quality with variable kernel size and high moisture content. Irregular yields are due to poor quality seeds, low rate of germination or low pest resistance.

A malt barley development programme, called 'Saanjhi Unnati' (SU), was created, with government agencies playing a significant role. The SU programme strives to build a long-term reliable source of locally grown malt barley and to test new strains of barley bringing better yield and quality. A key component of the programme was the creation of SU centres. These provide certified seeds, agricultural skills training, procurement services and other support.

SABMiller acts as the main coordinator and monitors the project. Cargill assists in all operational aspects, manages the SU centres through franchisees, sells seed and trains farmers. The Morarka Foundation facilitates social mobilization and interaction between local communities, SU management and operations staff. Rajasthan approves licences and provides infrastructure and support for the programme. The government also promotes the SU concept through its extension organizations, for example, by raising awareness of the benefits of certified seeds.

When the programme started, a publicity campaign was launched to raise the awareness among 20,000 local barley farmers, using a variety of communication tools including 'jeep campaigns', farmer meetings and leafleting. This led to a recruitment drive inviting farmers to join the SU programme. Participating farmers bought certified seeds from the SU centres and received access to personalized and group extension services.

Programme costs are estimated at between $92,000 and $156,000 per year for the first 3 years. The programme is not expected to become self-sustaining in the near term, and may need continued investment for an additional 3 years. In 2008, the 12 SU centres were supporting over 6000 farmers. The centres have distributed 500 t of certified seeds and expect to procure 16,000 t of barley in the first season. Around 60% of farmers who took the seeds sold barley back to the centres.

In the latest harvest, the yield was over 20% higher, although it is difficult to estimate the share that can be directly attributed to SU. Nevertheless, over a 2-year period, SU districts achieved an average yield increase of 33.3%, while the average for Rajasthan was 14.2%. Average thousand corn weight (TCW), a measure of malt extract, has gone up from 37 to 43.5 g since the SU programme was introduced.

Farmers were initially reluctant to accept seeds provided by SU. The government played a key role by publicizing the programme through extension activities and by lending credibility to it, assuaging farmers' doubts. Because farmers are price-sensitive, it proved important to ensure that the SU price consistently beat the open market price.

In the next 5 years, seeds from the seed development programme will be progressively distributed to farmers. Over the same time, SABMiller India expects to achieve 50% of total barley procurement from the SU programme, up from 10% in 2007.

Case 2: Nestlé facilitating the production of biogas in China

Nestlé facilitated an innovative approach for its milk suppliers in Shuangcheng in north-eastern China during the period 2004–2007, to convert the manure from dairy cattle into biogas to be used for cooking, heating and electricity generation. From the very beginning, the initiative was endorsed and strongly supported by the local authorities, and received further support from the central government. Chinese authorities are now replicating and scaling up the approach nationwide and extending it to pig farms.

While the strong demand for milk in China has had important economic benefit for the dairy farmers, it has also produced a new environmental challenge: how to manage the manure generated by dairy cows. Traditionally, most farmers compost manure outside the farm wall and apply it to the fields in spring and autumn. With the increasing number of cows, the handling needed revision.

The China Biogas project is part of the wider Sustainable Agriculture Initiative Nestlé (SAIN) launched in 2000 to optimize the supply chain from 'farm to factory' by improving efficiency, better managing risks and supporting sustainable agriculture. In China, Nestlé aspired to help its suppliers store manure in an appropriate manner, thereby reducing the risk of contaminating groundwater. Equally important, it wanted to create value for the farmers by converting manure into biogas for domestic usage, replacing the maize stems used by many families as domestic fuel.

Nestlé initially identified cheap, adequately sized biogas digesters as a possible solution. A key step was then to identify and convince a local 'champion' that would be willing to carry this project on the government side. In 2004 and 2005, trials of a small manure storage system with a capacity of $8\,m^3$ proved to work even during the cold winter of $-30°C$. The initiative was endorsed and strongly supported by the local authorities, and received further support from the central government. In order to demonstrate this technology, the Chinese authorities agreed that 74 Nestlé demonstration farms would be equipped with such a unit, and on-site training would take place during and after construction. Nestlé has played an active 'facilitator' role, creating awareness and stimulating demand from farmers to have such systems, and linking these farmers with construction teams.

The scale-up of the initiative was realized through the engagement of several partners, primarily from the public sector. The local government financed the establishment of five construction teams to replicate the approach. The central government provided the technical drawings, sealant, pipes, filters, cooking stoves and further equipment. In return, farmers were asked for a one-off payment to cover the construction material (cement, bricks).

Currently, there are 4000 small biogas digesters ($8\,m^3$) in Shuangcheng on an equal number of farms generating energy for the cooking stove. The new method of cooking has led to cost savings for farmers' families and better handling of manure. The annual cost for using methane gas bottles for an average household is RMB 400, which is just equivalent to the total investment of the biogas digester. Farm manure is now more effectively used and the remains are used as fertilizers. This decreases the need for synthetic fertilizers, which require non-renewable energy for their production.

The cooperation between Nestlé and Chinese authorities is a good example of how such projects can truly realize their potential. Biogas digesters are now available to dairy and pig farmers across China. Between now and 2010, additional biogas digesters will be constructed at a rate of 1000 digesters per year by each of five construction teams.

Case 3: Syngenta recovering water springs in Brazil

With its philanthropic project 'Agua Viva', Syngenta has contributed to the recovery of over 2100 water heads in Brazil. In collaboration with local cooperatives, Syngenta is providing technical assistance and know-how to farmers to recover their endangered and polluted water heads. As a result, springs are sanitized and protected, the quality of water increases and the water supply is secured for local communities.

Deforestation, agriculture and population growth have contributed to lowering freshwater reserves in most agriculture-based communities in Brazil. In this context, it is particularly important to reduce freshwater pollution. Farmers within the project area use fountainheads, which are natural springs, as sources of fresh water, both for domestic use and for important economic activities such as poultry, swine and cattle ranching. However, such fountainheads, if not properly looked after, are prone to several types of contamination.

'Agua Viva' started in 2004 in southern Brazil in collaboration with the farmer cooperative Coopavel. The project was run on the ground by a specialized technician with support from a water analysis laboratory. Fountainheads can be recuperated with simple and economical practices: all that is needed is rocks, a few kilograms of cement, plastic tarpaulin, short sections of PVC tubing and a little manual labour. However, the process requires efficient technical and environmental knowledge. The water spring recovery process takes typically 6 hours at a total cost of US$195 per fountainhead.

In the next phase, the approach was scaled up in several waves. 'Project multipliers' were trained to support 33 municipalities in the semi-arid north-eastern region of Alagoas, in partnership with the local Carpil Cooperative. In 2007, a major collaboration was agreed with the coffee cooperative in Guaxupè (Minas Gerais).

Recently, the project has evolved in a new direction as fountainheads are now registered in collaboration with the local government. In a recent pilot 'Agua Viva' is supporting the registration of water springs through GPS satellite positioning. The project is evolving into a partnership with the local government. This latest development truly aims to secure the water mines for the future generations and will help the authorities enforce the laws.

The project has won public acknowledgement on several occasions, including two awards: for being the 'best environmental project' developed in Brazil in 2004, under the 'Innovation' category awarded by ANDEF (Brazilian Crop Life); and as an 'outstanding environmental project', awarded by the Brazil–German Chamber of Commerce in 2005.

Since 2004, 2114 fountainheads have been recuperated, benefiting over 2700 families and their livestock. The improved water quality and sustainability of the springs generate rapid and direct results in the area of health and well-being. This is underlined by the large interest and demand in the targeted regions: waiting lists have formed to be part of the initiative.

The 'Agua Viva' project was able to make a difference with limited funds through a combination of technological know-how and close collaboration with local actors. This approach has not only allowed for contact to be established with recipient farmers and credibility to be built, but also for the sustainability of the project through the recruitment of a local 'champion' to take responsibility for e.g. the regular light maintenance required.

Case 4: Reliance India sourcing directly from local farmers

Reliance Retail is directly sourcing fresh agricultural produce from thousands of farmers through collection centres (CCs). The company is providing a guaranteed market for the farmers' produce, reducing transaction costs and training them in better and more sustainable farming practices. This initiative has resulted in increased income and upgraded skills for the farmers, a reduction in spoilage of produce by up to 35% and better quality products for Reliance retail stores.

Until recently, under regulations in place since the 1960s, all produce had to pass through state markets (mandis) with the intention to ensure 'fair' prices for the farmers and to reduce hoarding of agricultural produce during food shortages. While this regulation was phased out a few years ago, most retailers continued to procure their supplies using mandis.

In November 2006, Reliance Industries Limited entered the retail business with 'Reliance Fresh' stores. A key challenge was to source a supply of fresh produce for its stores of sufficient quality. Reliance created rural CCs across India where farmers could sell their fresh crops, fruits and vegetables locally. Through CCs, Reliance is pursuing the strategy of building an integrated business model that sources agricultural produce directly from the small farmers in Indian villages. To make this business model work and improve the quality of produce sourced through the initiative, the company is building human capital through education.

Before establishing a CC, Reliance raises awareness of its activities in surrounding villages and provides training on how to cultivate the desired products. Reliance also distributes seeds or saplings from Reliance-owned nurseries. CCs maintain farmer contact lists and call on them when they are looking to procure specific crops. By 2007 Reliance was operating around 160 CCs across India, with each CC buying produce from villages within a 15 km radius. During the harvest season, farmers bring their produce to the CC every day. Their product is weighed electronically and farmers are quoted a price matching the mandi's price. Reliance guarantees that it will purchase all produce delivered (which mandis do not) and the handling fees are approximately 50% lower than the mandis'. For higher-quality produce, farmers are offered a price premium. CCs load produce into plastic crates and label them to allow traceability. Reliance uses its own fleet of trucks to transfer fresh produce daily from CCs to regional processing and distribution centres, which then distribute it to Reliance Fresh stores. The produce is delivered on time and the spoilage is reduced to less than 5%.

Reliance is putting mechanisms in place, such as farmer interviews, to measure the impact of its direct farmer procurement in a systematic fashion. While the results are by no means exhaustive, they do provide evidence of the positive impact of the local sourcing efforts. Farmers receive a fair weighting and price and on-the-spot cash for their produce, and have improved the quality and efficiency of their production due to the price premium offered for higher-quality produce and the training offered by the CCs on efficient and sustainable agriculture techniques. More aware of differences in the quality of their produce, farmers are investing in seeds and inputs to grow higher-quality produce.

CCs demonstrate a step towards sharing value with smallholders by sourcing directly from the farmer communities. They allowed Reliance to gain relevant experience working with farmers, in view of implementing a more ambitious Rural Business Hubs concept. The exclusion from the supply chain of traditional intermediaries, however, led to protests, raising the question of how to evolve potentially obsolete supply chain structures.

Case 5: Unilever and partners building up a supply chain for Allanblackia (AB) oil

In Tanzania, Unilever was one of several organizations that partnered to establish a locally owned supply chain for AB oil, a new raw material to be used in margarines and spreads. Called Project Novella, the initiative is increasing income for farmers cultivating AB trees, generating jobs in the supply chain and preserving the biodiversity of the region.

Learning from an improved understanding of the impact of producing palm oil, Unilever is expecting that AB oil will also be used in the production of margarine. The AB tree is commonly found in parts of West, Central and East Africa. Unilever has conducted extensive research on the properties of AB oil and established new applications in manufacturing spreads and soaps. The oil's unique properties allow lower saturated fat versions of margarine. While AB oil is used by the local population, it is not produced commercially. Unilever's challenge is to source the volumes required to manufacture products from the new crop. The company has partnered with several NGOs, international organizations and government agencies in Tanzania to commercially produce food-grade AB oil.

Project Novella's main objective is to set up a new supply chain, sourcing AB seeds from local, community-based farmers. The project also aims to build human capital by training the farmers to produce high-quality AB in an environmentally sustainable manner. Multiple local NGOs, international organizations and government agencies have partnered with Unilever to implement the project. Challenges include local capacity building, social mobilization and change, research and technical training in sustainable domestication and plant propagation. Additionally, Unilever believes that, to make the model viable in the long run, the supply chain has to be owned by local farmers, small enterprises and communities.

Over the 4 years that Project Novella has been operating, the collaboration has made significant progress in creating a five-step supply chain. Farmers and groups in the villages are engaged in collecting AB seeds from their farms, drying the seeds and weighing and selling the product to collection centres throughout the harvest season. Seeds are then taken to a local crusher, who received support from Novella to upgrade his factory to produce food-grade oil. The AB oil is then transported to Europe. Unilever has guaranteed farmers a fixed price per kilogram of seeds, and has pledged to pay an attractive premium price for AB oil until the full economies of scale take effect in or before 2012. NGOs have raised awareness of the economic benefits of growing trees, and trained farmers and village associations on domestication. They have also helped villagers register as economic groups, giving them access to training and the opportunity to sell their produce to Unilever. As of 2006, 6000 farmers were involved in the programme; the average earnings per farmer per year increased from £30 to £70; 45 full-time jobs managing the buying centres were created and 650t of AB seeds produced. By the end of 2007, over 20,000 AB trees had been planted, most produced by rural, community-owned nurseries. By 2016, more than 25,000 farmers should be able to earn more than £200 per year farming AB, in addition to their other economic activities.

While the project has been successful, a growing supply chain will have to be established and made sustainable. Local ownership, active capacity building of the farmers, investment in research capabilities and improvement of business practices are some of the factors that are expected to contribute to the supply chain's sustainability and viability. The project is currently being replicated by Novella teams in Ghana and Nigeria. Moving forward will require attracting skilled people who can guide the process, as well as the funding to finance tree production and training for farmers on how to integrate AB into their current farm activities.

Case 6: Unilever committing to purchase certified tea for its key brands

Unilever is working with the Rainforest Alliance to certify all tea purchased for its Lipton brand by 2015. As the world's largest tea buyer, Unilever intends to send growers a clear signal in favour of sustainable tea production, emphasizing a long-term business model based on quality rather than on quantity in a market characterized by oversupply.

The tea sector is suffering from oversupply, with prices having dropped by 35% in 25 years. This situation has stabilized in the last few years; however, there are barriers to tea producers exiting the industry. Tea bushes live for over 100 years, and many farmers in developing countries have little knowledge of or access to alternative sources of income.

Unilever buys around 300,000 t of tea per year, or 12% of the world tea supply. Its own tea estates in Kenya supply 20% of the company's needs, while the rest is bought from other suppliers: around 750 big estates and one million small farms. Overall, two million people are directly involved in Unilever's tea supply chain.

Unilever did not want to simply launch a niche product, but to move the whole market. Certification was identified as the way to roll out the standards to the rest of the supply chain and credibly tell the story to the consumer. After discussing with three potential partners, the Rainforest Alliance was chosen by Unilever for conducting the external tea certification. The Alliance focuses on farm management aspects and its approach is comprehensive and applicable both to large estates and to smallholders and foresees collaboration with local organizations. The tea certification allows a market-driven increase in tea prices of 10–15%. The intention to certify tea was announced by Unilever in May 2007; in July, its Kenyan estate was certified, as well as one supplier.

The Rainforest Alliance developed tea certification criteria building upon non-crop-specific aspects (about 80% of all the criteria) and adding crop-specific aspects (remaining 20%). In addition, a country-specific component is taken into account by collaborating with local stakeholders. Every estate the Rainforest Alliance certifies is visited once a year to check that it is maintaining standards.

Unilever expects larger estates to show interest and take steps towards certification. For smallholders, typically delivering 'premium quality tea', a PPP was set up to promote and accelerate the movement towards sustainable tea production. Smallholders will receive support from the Rainforest Alliance, which will train them in farm management, prior to initial inspection. Over time, it is expected that they will save on input factors (e.g. pesticides). DfID is providing GBP 500,000 financing for this part of the tea certification. The Kenyan Tea Development Agency (KTDA), which represents the growers, is collaborating with local farmers. In Kenya, around 430,000 smallholders are expected to become part of the initiative, including smallholders not supplying Unilever.

Historically, tea was not in the focus of the Kenyan authorities, since this sector was strong in the country. However, recent developments suggest that more attention is being turned towards the tea industry. The public sector is looking at various ways to promote the tea sector through supporting upcoming schemes, like the Rainforest Alliance certification. One area where support is needed is in the diversification of income for tea growers. The government and multilateral donors might want to think about how to support those producers who cannot meet the standards and provide them opportunities to transition to other areas of economic activity.

Case 7: Grameen Danone setting up yogurt production in Bangladesh

---

The traditional approach to developing an international business in emerging markets is to target segments in which a company's product is likely to be competitive. This strategy typically limits developing country markets to the wealthiest customers. Following another approach, Danone Grameen is rethinking the entire value chain for yogurt production and marketing so that it employs and serves the poorest segments of society.

Danone has partnered with Grameen Bangladesh to deliver a quality, nutritious product to the 'base of the pyramid'. Its three objectives are to provide children suffering from nutritional deficiencies with a low-priced yogurt adapted to their nutritional needs, to create jobs around an economically viable and scaleable business model, and to preserve the environment.

The Global Alliance for Improved Nutrition (GAIN) is a critical actor in the joint venture, validating the benefits of the nutrient-enhanced yogurt for children.

For Danone, creating Grameen Danone Foods Ltd is consistent with its strategy to deliver health through nutrition. For Grameen, the joint venture is a natural addition to its existing portfolio of for-profit and non-profit enterprises serving the poor in Bangladesh. The Danone Grameen joint venture leverages Grameen's extensive reach and credibility in rural communities. The partnership provides GAIN with an avenue to pursue its mission of providing essential micronutrients to populations in need, leveraging both public and private food distribution channels.

Danone Grameen is addressing all essential components of the value chain, from agricultural production of milk to yogurt manufacturing and product distribution. Upstream, the objective is to promote local milk supply, particularly through the development of microfarms. Danone is using its experience to raise quality and productivity standards. The company is building local agronomic capacity by training local NGOs that can then work with farmers. Danone is also helping to establish a supply chain that prevents deterioration of milk in Bangladesh's hot climate.

At the centre of the value chain, Danone has designed a completely new factory set-up, which favours employment over costly technologies, without jeopardizing product quality. Downstream, the joint venture is relying on the extensive network of 'Grameen Ladies' for door-to-door sales along existing local retail channels. The goal is to keep distribution to a radius of 30 km around the factory to minimize environmental impacts associated with transport. However, the model introduces new challenges for the Grameen Ladies as yogurt distribution differs considerably from micro-lending or provision of telecom services, their existing activities.

Danone Grameen has invested significant effort in creating a product suitable for the market. The 'Shakti doi' (energy) yogurt is made of fresh milk and sugar and is enhanced with micronutrients (including e.g. vitamins, iron, iodine, zinc). It is cheaper than other yogurts and conceived to be in line with what low-income people can afford.

While much of the project has yet to unfold, the partners predict that over 1000 livestock and distribution jobs will be created (30 factory jobs were created in the pilot factory). Thus, employment impact will occur principally upstream and downstream in the value chain. While the venture is off to a promising start, the viability of the model has yet to be established; it is particularly challenging to operate a full value chain that relies solely on an extremely low-cost product. There is very little additional economic value to be captured in the market to finance the substantial investments required. The recently rising milk prices are further putting the business model under pressure.

8 Annex: Agriculture for Development – Implications for Agro-industries*

ALAIN DE JANVRY

Professor, Agriculture and Resource Economics, University of California, Berkeley, USA

Introduction

Every year the World Bank prepares the *World Development Report* (WDR) and, significantly, in 2008 the WDR focused on agriculture. The title selected for the report, *Agriculture for Development*, suggests that agriculture should be seen not only as a sector of economic activity but also as a means for the promotion of human welfare, food security, poverty reduction and environmental management. The WDR was developed in a highly collaborative fashion, including inputs from agencies such as FAO, IFAD and UNIDO, and involving consultations with many governments and organizations. It thus represents a broad understanding of the state of the art on the issues discussed. This chapter revisits some of the conclusions of the Report, with emphasis on the interface between agriculture and agro-industries in the process of development. Drawing from the Report, the focus of the chapter is on agro-industries and on what can be done to promote agro-industries within the context of agricultural development.

In 2008, agriculture was in the headlines, if for the wrong reasons. A cover of the magazine *The Economist*, for instance, gave prominence to the issue of rising food prices and the end of cheap food. Another issue of the same publication called attention to the rising disparities between urban and rural incomes and the political tensions this phenomenon created.

There are many reasons why agriculture has been in the headlines. It is enough to open newspapers every day to see that there is mounting concern with the issue of sharply rising food prices and with what this implies for food insecurity and the rise of hunger. Rising rural–urban income disparities are a source of political tensions. India, China and many of the rapidly growing countries find themselves with a large rural population being left behind and demand-

---

* Transcription of Alain de Janvry's keynote speech on 'Agriculture for Development–Implications for Agro-industries' delivered at the Global Agro-industries Forum, New Delhi, India, April 2008.

ing greater attention. The Millennium Development Goals (MDGs) will not be met in many parts of the world. Because 75% of the world's poor live in rural areas, meeting the MDGs inevitably requires giving greater attention to agriculture as a source of income for the poor. Threats to the survival of the family farm and excessively rapid migration to the cities, leading to overcrowding and unemployment, are also important concerns. New demands on agriculture for better-quality foods, rapidly rising consumer demand for animal and fish products and concern about the environment all create visibility for agriculture in development. Some recent health epidemics have been closely related to issues of water and proximity to animals, creating important links between agriculture and health. Finally, and very importantly, there is the issue of climate change, to which agriculture contributes importantly and from which agriculture already suffers, with particular impact on the poorest rural inhabitants.

Agriculture has been in the headlines and, as a consequence, it is timely to talk about how to make the most of agriculture for development. A particular concern is how countries can benefit from the links between agriculture and agro-industrial development and how agro-industry can contribute to economic development.

It is instructive to note how the WDR approached the issue of focusing on agriculture as a development driver. One basic concern was that of heterogeneity of conditions: since the world is vast and diverse, there are different ways of looking at the role of agriculture for development in different parts of the globe. Another concern was that there are many pathways out of poverty, some of which include agro-industry as an important element. A general conclusion, however, is that the potential that agriculture has to contribute to development has been vastly underused, especially in the last 25 years. Yet there are opportunities to do better and, as a consequence, we must identify those opportunities, confront the challenges associated with them and then ask ourselves: what are the policy entry points that can mobilize this potential in the context of the new opportunities and the new challenges? What are the conditions for success and what can we do to make it happen? Will the world go back to business as usual, continuing to neglect agriculture as an instrument for development? Or can agriculture be seriously used for development and make a real difference for growth in low-income countries, poverty reduction and greater environmental sustainability? It is a matter of urgency now: with agriculture in the headlines, there is no time to waste in finding out how to focus on agriculture and agro-industry more effectively as engines for development.

How Has Agriculture Served the Cause of Development?

To understand the roles of agriculture in development, it is necessary to look at the structural transformations as illustrated by the relationship between gross domestic product (GDP) per capita and the share of agriculture and agribusiness in GDP (Figure 1).

The horizontal axis of the graph presents GDP per capita and the vertical axis presents the shares of GDP for particular sectors. We know that the importance

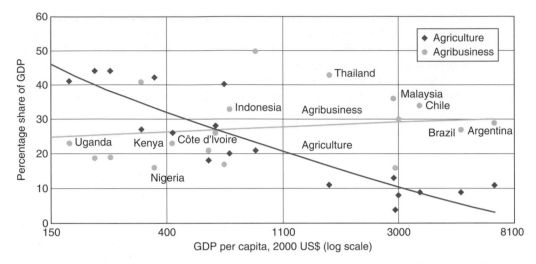

Figure 1. The structural transformation: a declining share of agriculture in GDP, but a high and rising share of agribusiness as per capita income rises.

of agriculture declines as GDP per capita increases from, say, 50% at low levels of income to approximately 5% at high levels of income. But the interesting aspect to notice is the trajectory for agribusiness (agro-industry and related services). What can be seen in Figure 1 is that the share of agribusiness is not declining – in fact, it is rising, tending to decline only later on, at higher levels of GDP per capita. As incomes rise, agribusiness becomes much more important than agriculture. Obviously, other sectors of industry are going to emerge, but the agribusiness sector, including all the services and activities linked to agriculture, is typically a very important part of the emergence of industry. Indeed, as shown in Figure 2, the

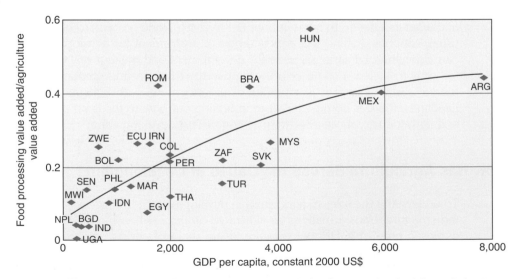

Figure 2. The structural transformation: value added in food processing is rising relative to value added in agriculture.

share of value added in food processing in relation to the share of value added in agriculture is of increasing importance as GDP per capita rises. Hence, it is important to think of agriculture not as a declining industry, but as supporting the emergence and growth of an industrial sector.

Acknowledging the fact that the world is quite heterogeneous, the WDR utilizes a country typology, as depicted in Figure 3. On the horizontal axis, the share of total poverty located in rural areas is presented. This share can be very high in countries like China, where about 95% of total poverty is found in rural areas, or in India, where it reaches some 80%. On the vertical axis, we have the share of growth attributed to the agriculture sector during the last 25 years. This framework gives us three categories of countries, each with a different policy agenda in using agriculture for development. The first category is the agriculture-based countries. These are largely the sub-Saharan African countries, where agriculture is an important source of growth and most of the poverty is rural, with poor households dependent on agriculture. For these countries, the main policy problem is to accelerate agricultural growth. The second category is the transforming countries, where growth has accelerated in other sectors of the economy and agriculture is no longer a major source of GDP growth, but most of the poverty remains rural. This includes countries such as India, China, Morocco and Indonesia. For these countries, the main policy problem is the rising income disparity between rural and urban sectors, and the persistence of deeply entrenched rural poverty. The third category is the urbanized countries, where agriculture makes a low contribution to GDP growth (although the share of agribusiness is typically larger), but where some 30–40% of poverty remains rural. For these

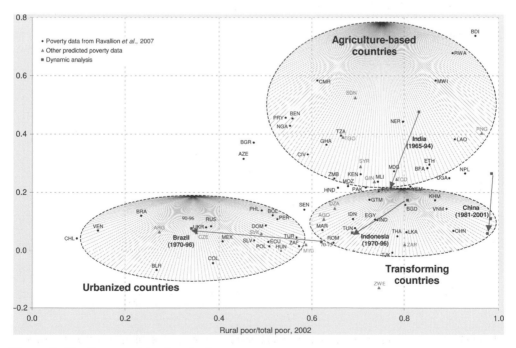

Figure 3. The three worlds of agriculture.

countries, the main policy issue is that of social inclusion of smallholders into the set of competitive farms. Do these countries want an agriculture that leaves room for a competitive sector of family farms, or an agriculture with mainly large farms and a labour market that becomes the instrument for poverty reduction?

There are three functions that agriculture can fulfil for development. First, agriculture can be the lead sector to trigger economic growth, particularly in the agriculture-based countries. Second, it is an important source of livelihoods. As we have seen, 75% of the world's poverty is rural. For most of the world's poor, agriculture is the main source of livelihoods and hence it has to be mobilized for that purpose. Third, agriculture is a very large user of natural resources and the way it uses resources has global impacts. Since global impacts have an important influence on agriculture, the links between agriculture and the environment have to be taken into account very closely.

There are many success stories in terms of these three functions. Agriculture-based countries for which agriculture is a source of growth include most of the sub-Saharan African countries, but also Paraguay and many Central American and Caribbean countries. Agriculture, in most situations, is the sector that has greater competitive advantage. Since agro-industry is a very significant way of adding value to agricultural products, it offers an important way to transit towards industrialization, whereby countries industrialize based on competitive advantage in agrifood processing and value addition and then increasingly move away from this source of competitive advantage by creating new sources that lead to industrial development, with a declining share of GDP coming from agriculture. This has happened, for example, in Chile and Malaysia, leading to acceleration of growth.

In terms of sources of livelihoods, there are important success stories as illustrated in Figure 4, which depicts the cases of India on the left and China on the right. The rising curve shows cereal yields, while the declining curve shows the rural poverty rate. Clearly, there is a high negative correlation between the two.

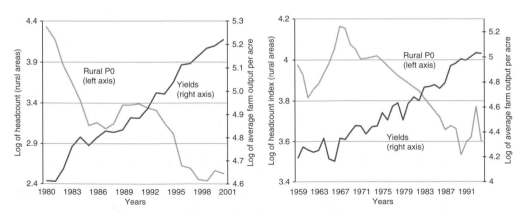

Figure 4. In India (left panel) and China (right panel): cereal yields and rural poverty rates.

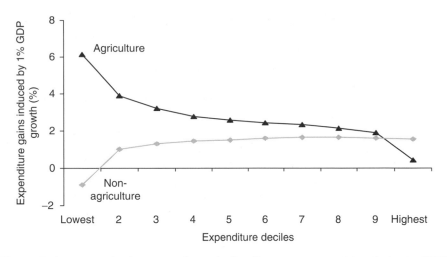

Figure 5. Income gains by expenditure deciles (from poorest to richest) due to GDP growth originating in agriculture (triangles) and non-agriculture (lozenges).

These cases can be contrasted with what has been happening in other parts of the world, such as sub-Saharan Africa, where yields remained static and poverty persistent.

To further this argument, it is helpful to calculate how much income gain can be obtained, for different segments of the income distribution curve, from GDP growth originating in agriculture or in the rest of the economy. In Figure 5, the ten deciles of the distribution of income are shown on the horizontal axis, calculated from aggregated data from 42 countries (WDR 2008). The curve with triangles shows the income gain for each of those deciles coming from GDP growth originating in agriculture versus the curve with lozenges when GDP growth originates in the rest of the economy. It can be seen that, for the 50% poorest, the gains in income are much larger (on the order of 2–3 times) when growth originates from agriculture than when it originates from the other sectors of the economy.

This has significant implications for world poverty. Since 75% of world poverty is rural and since agriculture is the main source of income for these poor, then income derived from agriculture and agro-industry is a very effective means to reduce poverty. Moreover, as we have seen, agro-industry does offer a path towards industrial growth.

As can be seen in Figure 6, there exist in fact a number of pathways out of poverty. A policy agenda can thus be derived as to how to use agriculture and agro-industry as instruments to promote development.

The framework presented in Figure 6 shows that the main pathways out of poverty are via farming, the rural labour market and migration. Farming households consist of commercial smallholders, households that are in transition from subsistence farming to commercialization and the large subsistence farming economy. Each of these three groups requires different policy interventions, or differentiated policy packages that can cater to their specific needs.

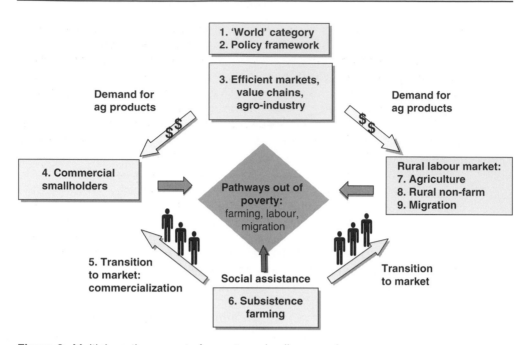

Figure 6. Multiple pathways out of poverty and policy agenda.

The rural labour market has two interrelated components. On the one hand, there is the agriculture labour market, which can be a way out of poverty, but also often a way of remaining in poverty. We know that agriculture labour markets tend to be quite harsh, informal, seasonal and often involve extensive participation by children. Since many of the poorest people in the world depend on agricultural employment for their subsistence, considerably more can be done to make the agriculture labour market into an instrument to move populations out of poverty.

On the other hand, and very importantly, there is the rural non-farm economy. This consists of a whole set of industries, ranging from small to medium and large firms that are engaged in forward and backward linkages with agriculture, catering to consumer demands that lead to agricultural incomes. Their location is quite often in rural areas, where they provide complementary sources of income and employment to rural households.

The rural non-farm economy is another very important instrument for poverty reduction. It has been widely observed that, where you have successful smallholder farming, it is usually complemented by a successful rural non-farm economy. In this context, rural households tend to be diversified, participating not only in farming activities, but also in employment and investment in the rural non-farm economy. Significantly, to a very large extent the non-farm economy is composed of agribusinesses, agro-industries and agro-services linked to agriculture.

Finally, there is the issue of migration. Population growth will force many to move out of agriculture, as there is not enough room for employment in agriculture for a rapidly growing labour force. Over time absolute employment in agriculture will inevitably decline. Yet, even though migration can be an important pathway out of poverty, it should be noted that in order to maximize the potential of its economic and social benefits, a precondition exists: there has to be a significant investment in the quality of the rural labour force in terms of health and education, so that the labour force can successfully migrate to other employment opportunities beyond agriculture.

To summarize, there are three pathways out of poverty – self-employment in farming, the rural labour market and migration. The farming-based pathway builds on smallholder competitiveness, i.e. a smallholder sector able to successfully compete on markets. Rural labour markets comprise agriculture and the rural non-farm economy, including, significantly, agro-industry. The migration pathway, on the other hand, requires preparedness. There is still a need to consider the large subsistence farming sector. Ways out of subsistence farming include two major options. First, farmers can become commercial smallholders via increased commercialization, increased access to assets and increased capacities for entrepreneurship. Second, farmers can move into the labour markets, but this requires added skills, especially if a household member is to enter the rural non-farm economy.

Supporting all elements considered in the framework depicted in Figure 6 are efficient markets, value chains, agro-industries and the links to the general policy framework as well as which of the 'three worlds' a particular country belongs to, in accordance with the typology established earlier. These considerations provide nine entry points that can be used to set an agriculture-for-development agenda:

1. The 'world' category to which the country belongs.
2. The policy framework.
3. Efficient markets and value chains.
4. Commercial smallholders.
5. Commercialization from subsistence to market-oriented smallholders.
6. The whole subsistence economy and what can be done in terms of social assistance.
7. The rural labour market, which includes agriculture.
8. The rural non-farm economy with agro-industry, agro-business and agro-services.
9. Migration.

We can use these nine entry points to propose agendas as to how to approach and enhance agriculture and agribusiness for development in particular country settings.

Agriculture for Development and the Nature of Public Spending and Donor Support

It has been noted that 75% of the world's poor are rural. Yet only 4% of the official development assistance (ODA) and 4% of public spending in sub-Saharan Africa goes to agriculture. These figures suggest that there is a lack of

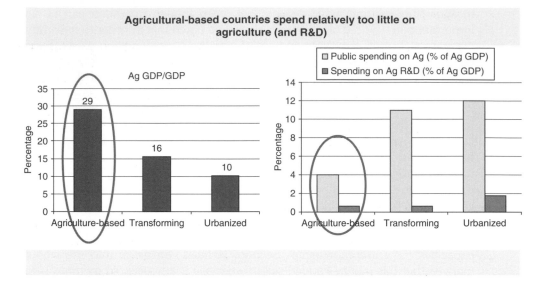

Figure 7. Public expenditure in agriculture is low in the agriculture-based countries, while the share of agriculture in GDP is high.

correspondence between the objective of poverty reduction and the instruments that are being used to reduce poverty, namely public investment and ODA. This is an important observation made in the WDR, which leads to the question as to what then can be done to have agriculture as a development engine.

The data analysed by the WDR show that there is indeed under-investment in agriculture The left-hand panel of Figure 7 shows that agriculture as a share of GDP is about 30% in the agriculture-based countries. None the less, it can be seen in the right-hand panel that public expenditure going to agriculture, as a percentage of GDP, is only 4% in these countries.

The New Partnership for Africa's Development (NEPAD) has set a goal of 10% for the share of the national budgets that should be allocated to agriculture. From 10% to 12% is what India was apportioning to agriculture at the time of the green revolution. By the same token, 10–12% is what China has been allocating for supporting its 'household responsibility system'. As such, the figure of 10–12% has become a useful benchmark as to how much public spending there should be as a share of the national budgets and we observe that many poor countries are way below this. Moreover, there is not only the issue of quantity, but also that of quality. Many countries have seen a drift in their public expenditures on agriculture towards subsidies and away from public investments. In India, for example, 75% of public expenditure goes to subsidies (electricity for pumping water, fertilizers, etc.), while only 25% goes to public investment. This has a huge opportunity cost in terms of possible alternative public spending targeting agriculture and rural people, such as research and development, infrastructure for agriculture or health and education for rural populations. Needless to say, decisions on the quality of public spending are a question of a country's political economy.

Agriculture for Development: the New Opportunities

Fortunately, there are new opportunities to make better use of agriculture for development than in the recent past. They include a 'new agriculture', with dynamic demand, high-value activities, non-traditional exports and ability to add value to commodities in agribusiness. The markets have changed, agricultural product prices are rising and opportunities for investment exist. Price transmission has to happen between markets and farmers, but clearly this is a time when we see unique incentives being provided to the agricultural sector. The big challenge will be to make investments in agriculture favourable, not only for growth, but also for poverty reduction and for environmental sustainability. New opportunities also come from numerous institutional and technological innovations. India has been a leader in this respect, using information technology for extension, linking farmers to markets and for information provision about innovations in terms of use of new seeds and pest control. Finally, and very significantly in terms of new opportunities, we now face an agricultural sector that is quite different from, say, 25 years ago, the last time that the World Bank focused on agriculture in the *World Development Report*. The private sector has now a very important role to play, not only in terms of agribusiness and value addition, but also through direct engagement in agricultural production. Producer organizations play an increasingly important role for service and advocacy. Despite the limitations to deliver what they ideally should to their membership, producer organizations are rapidly emerging and consolidating.

The new actors and the ways of doing business in agriculture call for a new role for the state and for new ways for governments to become engaged with the private sector and with civil society organizations in different forms of partnerships. Public–private partnerships (PPP), in which a more proactive state engages with the private sector and civil society organizations in promoting new initiatives, offer plenty of interesting opportunities that have to be better explored and understood.

Some of the new opportunities offered by a dynamic demand for agrifoods are illustrated in Figure 8, with data from India. In the left-hand panel, it can be

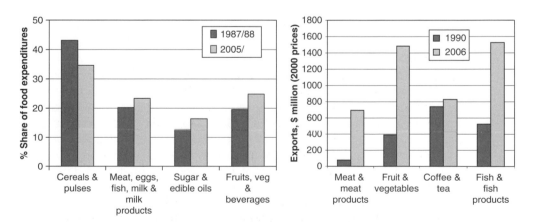

Figure 8. India: rising consumer demand and exports for high-value products.

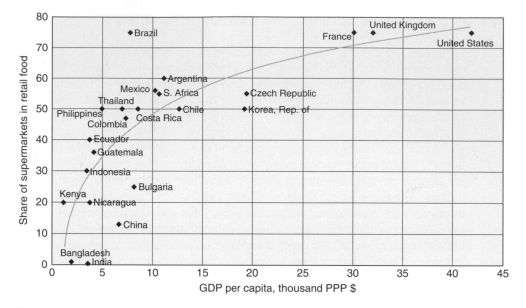

Figure 9. A rising share of supermarkets in retail food as income per capita rises.

seen that the share of cereals and pulses in consumer expenditures has been declining in the country, while the share of high-value products, meat and eggs, sugar and oils, and fruits and vegetables has been increasing. The graph in the right panel shows the significant increase in exports of high-value products from India.

Figure 9 shows the rise of supermarkets as food retailers as incomes rise, for a cross-section of countries. It can be seen that, increasingly, the nature of demand is being transformed by the emergence of the supermarket economy. Supermarkets tend to impose their own quality and safety standards, and hence they impose tremendous challenges on the way smallholder farmers need to organize in order to deliver the kinds of produce that supermarkets want to put on their shelves. The rise of supermarkets represents both opportunity and a challenge. While the potential for smallholder producers to capture rapidly expanding markets exists, there is also a risk that they might be cut off from access to their traditional urban consumers, as supermarkets can also procure internationally or from larger-scale farmers.

The potential poverty reduction effects of non-traditional exports can be illustrated from a study of bean exports from Senegal to France (Figure 10). In the left panel, we can see the rapid rise of the total number of participants in export sales between 1991 and 2005. We also see two interesting curves, respectively representing the number of smallholders operating under contract farming agreements with exporters and the number of estate farm workers. Over time, the pattern has been one of smallholders being displaced by estate workers. A question can then be asked as to what high-value agro-exports can do to contribute to the goal of poverty reduction.

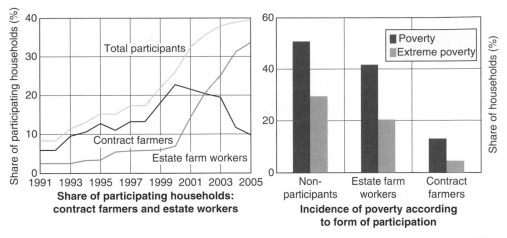

Figure 10. French bean exports and poverty in Senegal. (From Maertens and Swinnen, 2006.)

In the right-hand panel, the poverty rate is shown in the darker shaded bars, while the extreme poverty rate is represented in the lighter ones. Poverty rates are higher for the non-participants, suggesting that export markets contribute to raise farmer incomes. Comparing the two classes of producers, it can be seen that estate farm workers have gained from participation, but less so than their contract farmer counterparts. Increasingly, medium to large farms with hired workers tend to displace smaller farms, which can be competitive players if they are effectively organized. The key issue of concern thus becomes the role of the labour market and whether it can provide incomes to estate farm workers that can offer them a pathway out of poverty. This empirical evidence highlights the importance not only of competitiveness of a smallholder sector, but also of an agricultural labour market that can afford working conditions that allow people to move out of poverty.

Figure 11 shows the difference in employment across different countries between two types of farm enterprise: cereal production and high-value products (vegetables). The data are expressed by the number of labour days per hectare for cereals and vegetables, which are respectively shown by the light and dark shaded bars. The figure illustrates well the tremendous gain in employment that happens in the shift from cereals to vegetables.

High-value crops are quite effective in creating employment opportunities. To promote this type of agricultural enterprise, a better investment climate is needed that offers incentives to invest in high-value crops and comprises institutional innovations such as effective producer organizations, contract farming and PPP. Also important are improved sanitary and phytosanitary regulations. These are all areas for which the public sector has an important role to play, as a complement to what the private sector can do.

Other opportunities for a renewed focus in agriculture for development are offered by the recent phenomenon of rising commodity prices. Figure 12 shows that the price of wheat, for instance, has more than doubled in the last

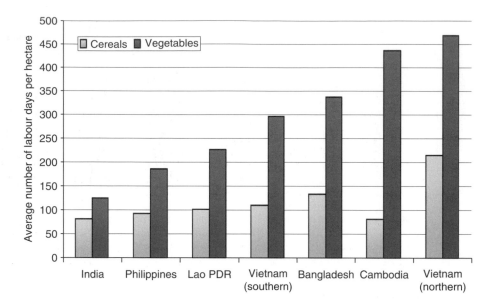

Figure 11. Labour intensity of vegetable production relative to cereals.

Figure 12. A sharp rise in the price of wheat.

3 years, creating both opportunities for producers with a marketed surplus, and major challenges for net buyers, which include not only the urban population and landless rural people, but also a surprisingly large share of smallholders.

Figure 13 shows the shares of total employment in the rural non-farm economy by type of economic activity in different regions. In the case of sub-Saharan Africa, significant employment comes from retailing. Next is South Asia, where the services and manufacturing sectors are much more important. This is where the agribusiness and agro-services sectors come into play, creating employment opportunities in the rural sector, which then complements what the farming sector can offer to rural households.

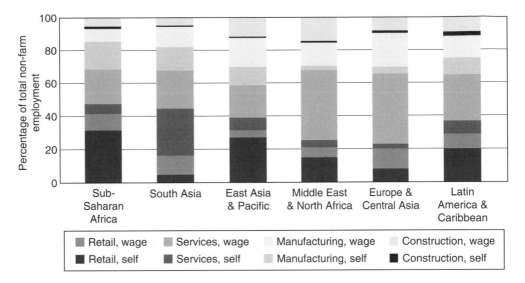

Figure 13. A diversified rural non-farm economy.

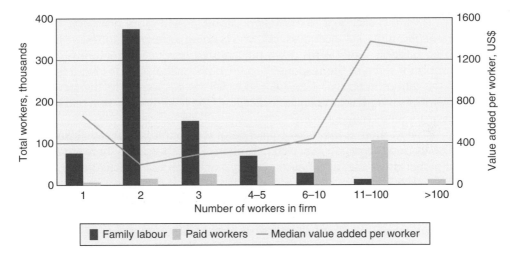

Figure 14. Employment and labour productivity in the rural non-farm economy by size of firm.

To characterize employment in the rural non-farm economy, Figure 14 shows the value added per worker in Indonesia for firms that range from very small (only one worker) to very large (more than 100 workers). The dark shaded bars represent employment of family labour. Hence, in the small firms are found mainly family workers but larger firms increasingly have paid workers.

However, the important observation to be inferred from the figure is the value added per worker across firm sizes. Firms with three or less employees tend to have very low labour productivity, which only increases in the medium-scale and larger firms. This has important implications for agro-industries. Smaller firms are, on average, really sources of hidden unemployment, while medium and larger firms are really the ones that are productive and able to deliver competitive levels of wages and better working conditions to their employees. As such, care must be taken when describing the rural non-farm economy and the opportunities it offers for poverty reduction, as the sector is highly heterogeneous, with many low-productivity activities existing in small firms.

Agriculture for Development: the New Challenges to Be Met

One of the biggest challenges facing the role of agriculture for development is the political economy of public expenditure. Quantitatively there is a need to increase public investment, while qualitatively there should be a move away from subsidies. Not all subsidies are necessarily bad, but many fail to support productivity gains.

A second challenge is that, in many countries, there is a considerable lack of definition with regard to the relative roles of the market and the state, with the conflicts and inefficiencies that this brings about. There is not one solution as to what should be done by the private sector and the market versus what should be done by the state. However, what is important is that the 'rules of the game' must be clear and that policies are followed consistently.

A third challenge is to make agricultural growth more pro-poor. There are tremendous market incentives, but who is going to respond to them? Is it going to be basically large farms using mechanized production methods, or will an important share of the response come from smallholders? Is investment going to be labour intensive? Can we reconcile the responses to the new market incentives with development objectives such as poverty reduction and improved environmental management?

Finally, there is the rising challenge of environmental sustainability – water scarcity, land degradation and climate change. Furthermore, there are a number of global issues that affect each country's ability to use agriculture for development. The Doha Round of trade negotiations must progress. Genetically modified organisms (GMOs) may offer opportunities that have not been fully explored in terms of smallholder farming and nutritional value. Biofuels can be important but, except for Brazil and a few similar countries, they are, at the moment, sufficiently efficient neither in terms of cost-effectiveness nor in terms of CO_2 savings. There is a need to invest in identification of a new generation of biofuels that can be more environmentally and economically viable. Subsidies to ethanol in the OECD countries are destabilizing the world food situation, with a high cost for the poor when they have irrelevant benefits for the rich.

Policy Entry Points in Having Agriculture and Agro-industry as Focuses for Development

To identify policy entry points in focusing on agriculture for development, let us go back to the nine-point agenda we considered when discussing the framework presented in Figure 6. The first step is to recognize that the 'worlds' of agriculture differ and that agriculture fulfils different functions in each 'world'. Countries need to identify which world they are in: agriculture-based, transforming or urbanized. Agriculture has different functions to fulfil in each of these categories. It can be a source of growth, it can address the problems of income disparities and poverty and it can be a means for economic inclusion, with smallholders achieving competitiveness and remaining as actors in the agrifood chains.

Second, policy options are different in a context of high food prices and of high-value products and agro-transformation. Clearly, the issue of supply response is back on the policy agenda. For the last 50 years, which witnessed steadily declining food prices, the focus of international development agencies was more on the consumer side, i.e. on promoting access to food and to sources of better nutrition. Renewed importance is now being attached to the supply side, for the simple reason that prices are rising and, as such, production has to increase. Countries are concerned with how to mobilize productivity gains through improved infrastructure and institutions that can support value addition in agriculture and agro-industry.

Urban consumers and the landless in rural areas are affected negatively by high food prices. However, most of the world's poor farmers are also net buyers of food and not net sellers. Hence, they are negatively affected by the rise in food prices. As a result, when there is a situation of rapidly rising food prices, we should not assume that all farmers will be positively affected. This is important, because they are typically very difficult to be reached by social assistance programmes, such as food aid or cash transfers.

For the third group of countries, issues related to market access, value chains and agro-industries become particularly relevant. Transaction costs are too high on many markets, particularly in sub-Saharan Africa. There is insufficient value addition in agribusiness. In many agro-industries, especially in the international market, there is a very rapidly rising degree of concentration, which is of concern in terms of market efficiency. One example in this regard is that of the coffee sector, which has been booming in terms of adding value, but where both the relative share and the absolute value of the market revenues remaining in producing countries have been decreasing over time. A second issue is smallholder access to agro-industry. Smallholders relate better to small and medium enterprises than they do to large enterprises. It is important to understand how to create incentives to allow small and medium enterprises to enter the industry and compete and in that respect the investment climate is a relevant consideration. Enabling conditions are needed, such as functioning markets, access to financial services and institutions for contracting and risk-sharing. The role of PPP to promote investment may be important. The state can be proactive in initiating PPP, in particular with small and medium enterprises in the agro-industrial sector, in order to facilitate entry and achieve competitiveness. With regard to the role of labour practices, the Chilean model is considered a successful approach to poverty reduction that is based on employment conditions in the agro-industrial sector, such as year-round employment achieved through product diversification. Regulations were enacted that do not discourage employment, protect workers and allow work conditions that are conducive to poverty reduction.

The next element on the nine-point agenda is smallholder competitiveness and here there are two important dimensions. One is that there is still a missing dimension to the green revolution, with Africa not having experienced a quantum jump in yields equivalent to that which occurred in Asia. The Asian experience with the green revolution is well known, but for it to be reproduced elsewhere requires careful adaptation of the Asian experience. Clearly, the conditions in Africa are different. Farming systems are much more heterogeneous and there are many more crops that have to be dealt with. Markets have to be put into place and institutions have to be created, all at the same time as the technology that can support the green revolution is provided. Second is the key role of a high-value revolution in agriculture and the fact that many countries are moving towards it, but quite often not as rapidly as they could be and often with limited smallholder participation. India is a case in point with tremendous potential to place itself in international markets of high-value commodities and to supply the emerging middle-income domestic consumer market, which is urban based and likely to be increasingly serviced by supermarkets. Yet progress is taking place slowly, due to lack of adequate infrastructure, poor supportive institutions and producer organizations that are poorly developed to allow smallholders to be part of the high-value revolution.

Next on the agenda is family farming in transition to commercialization, i.e. taking households out of subsistence farming and into the market. Greater attention is needed to subsistence farming, which requires different types of support in order to improve the resilience and the capacity of the farming systems to achieve food security.

The rural labour market is potentially a very important pathway out of poverty, because, as we have seen, many of the poorest rural people in the world are actually in the agricultural labour force. But these markets are often not properly regulated and not functioning adequately if they are to achieve poverty reduction.

The rural non-farm economy, presenting employment and investment opportunities, is a further element in the agenda. Here, there are basically three policy entry points for discussion. First is the rural investment climate in all of its dimensions, which needs to be carefully understood. Second is the new role of the state in terms of PPP. Third is the role of clusters and a territorial approach, which can coordinate public support and public–private investments at the level of specific regions. Under these circumstances agro-industries can develop competitive advantages that are based on the comparative advantages of agriculture and on cross-firm spillovers to enhance their competitiveness.

The final element in our list is investing in rural people. There has been massive under-investment in rural people in terms of health, education and gender equity, and it is no surprise that, as a consequence, as labour moves out of agriculture, it is unable to find good employment opportunities.

In sum, there are basically three main messages. The first is the tremendous importance and potential for the promotion of smallholder competitiveness in high-value activities, in addition to the more traditional focus on food grains. For competitiveness to be achieved, the development of proper institutions, infrastructure and organizations will be required The second is rural employment in high-value activities, or the role of the rural labour market and how wage labour will play a role in the transformation of agriculture. The third is value added in agro-industries and the need to explore, quite carefully, the conditions that can be put into place in order to favour the emergence of these enterprises, especially medium-size ones that are highly employment intensive, using labour in a way that is both remunerative and conducive to leading workers out of poverty.

How to Make It Happen? Conditions for Success

What is the likelihood that agriculture and agro-industry will now be used more effectively for development than they have been over the last 20 years? Basically there are four conditions that should hold in order for this to happen. The first is awareness: governments and development agencies need to better understand what agriculture can do for development. This has, to a large extent, been forgotten over the past 2 decades, when we shifted towards an urban-based development model, an industry-based model of development or towards high-tech services such as in India. And yet, although we see wealth creation, we also witness continued mass poverty in rural areas and the growth of rural–urban disparities, with an underuse of the potential that agriculture has for poverty reduction.

Second are the possibilities to invest public resources in agriculture competitively, not only for growth, but also for poverty reduction and sustainability. As we have seen, there are important new market, technological and institutional opportunities for agriculture, but these incentives to invest need to be reconciled with what they can do for development, namely poverty reduction and sustainability.

The last two conditions are the biggest challenges. The first is how do we invest successfully in agriculture as an engine for development? If governments and international development agencies need to focus on agriculture for development, how should they do it? With the food crisis creating opportunities for political support, the World Bank, the Bill and Melinda Gates Foundation, the Rockefeller Foundation, FAO, IFAD and many bilateral development agencies promoting investments in agriculture are then asking: what are the options? How do we invest successfully in agriculture for development? The old ways will not work: the present context is different, and the traditional approaches did not work for Africa. There is thus a need for new ideas, a need for experimenting with innovative ways of using agriculture for development, such as decentralization policies, participatory approaches, strengthening community roles, reinforcing the important roles of the public sector, promoting PPP, increasing the focus on women in agriculture, reshaping the investment climate, strengthening producer organizations, seeking value addition in agribusiness, and so on. Agriculture has tremendous potential to promote development, but it is not infallible.

The final condition is the need to develop capacities. In order for agriculture to be effectively used for development, we need to build capacities, especially in terms of entrepreneurship. It is not easy to invest in high-value activities; it is not simple to invest in small or medium agro-industries; it takes entrepreneurship, business skills and education, which have usually not been provided adequately by schools in the rural sector. Second, small farmers or small entrepreneurs will not succeed alone: they need to work with producer organizations; they need to cooperate in order to compete successfully; and hence they also need leadership and analytical and managerial capacities at the level of their organizations. Third, we have seen that the state has a very important role to play: you cannot have a successful agriculture-for-development programme without an active and supportive state. The investment climate will be managed by the state but, for this, the state has to have a good understanding of what the private sector needs in order to be successful in investing in agriculture and agro-industry. Governance for agriculture needs to be redefined and the role of ministries of agriculture needs to be rethought. They have usually been badly affected by structural adjustment: resources have declined and functions have been redefined, but we have not seen enough adjustments regarding the kinds of skills and organizational structure they should have. And, finally, there has to be support at the international level. No country can now succeed alone in using agriculture for development; there are too many international dimensions that have to be understood and supported. Thus, international support and coordination are essential and, for this, capacities have to be developed also at the level of international organizations.

References

Maertens, M. and Swinnen, F.M. (2006) *Trade, Standards and Poverty: Evidence from Senegal*. LICOS Centre for Institutions and Economic Performance and Department of Economics, University of Leuven.

World Bank (2007) World Development Report 2008: Agriculture for Development. Washington, DC.

Ravaillon, M., Chen, S. and Sangraula, P. (2007) New evidence on the urbanization of global poverty. Background paper for the world development report 2008. The World Bank, Washington, DC.

Index